T0336209

ANTISOCIAL MEDIA

POSTMILLENNIAL POP
General Editors: Karen Tongson and Henry Jenkins

# Antisocial Media

*Anxious Labor in the Digital Economy*

Greg Goldberg

NEW YORK UNIVERSITY PRESS
New York

NEW YORK UNIVERSITY PRESS
New York
www.nyupress.org

© 2018 by New York University
All rights reserved

References to Internet websites (URLs) were accurate at the time of writing. Neither the author nor New York University Press is responsible for URLs that may have expired or changed since the manuscript was prepared.

Library of Congress Cataloging-in-Publication Data
Names: Goldberg, Greg, author.
Title: Antisocial media : anxious labor in the digital economy / Greg Goldberg.
Description: New York : New York University Press, [2017] | Series: Postmillennial pop | Includes bibliographical references and index.
Identifiers: LCCN 2017013006 | ISBN 9781479829989 (cl : alk. paper) | ISBN 9781479821907 (pb : alk. paper)
Subjects: LCSH: Information technology—Economic aspects. | Digital media—Economic aspects. | Labor.
Classification: LCC HC79.I55 G64 2017 | DDC 306.3/6—dc23
LC record available at https://lccn.loc.gov/2017013006

New York University Press books are printed on acid-free paper, and their binding materials are chosen for strength and durability. We strive to use environmentally responsible suppliers and materials to the greatest extent possible in publishing our books.

Manufactured in the United States of America

10 9 8 7 6 5 4 3 2 1

Also available as an ebook

*For my parents,*

*Isabel and Mark Goldberg*

# CONTENTS

# ACKNOWLEDGMENTS

A number of people have supported me through the process of writing, and have inspired me with their honesty, courage, compassion, and humor. Given the argument of this book, I'm not sure I should say that I am grateful to them, but I can easily say that their companionship has meant a lot to me.

Patricia Clough was a formative influence. Like many of her students, I have tried to emulate her anti-disciplinarity, political savvy, facility with social theory, and, perhaps most of all, her ability to *read* a text. It was a pleasure to share Patricia with her other students—my partners-in-crime—at the City University of New York Graduate Center. Craig Willse and Rachel Schiff were particularly patient, generous and, most importantly, so much fun.

I wrote this book while an assistant professor in the sociology department at Wesleyan University. I feel lucky to have found a welcoming home, with colleagues and students who keep me on my toes and are game for whatever I throw their way. I am also lucky to have worked with New York University Press on this book: thanks to Lisha Nadkarni and the Postmillennial Pop series editors, Karen Tongson and Henry Jenkins, for their support, and to the manuscript reviewers for their feedback. In particular, Lucas Hilderbrand's thoughtful questions and comments were instrumental in making this a much better book.

While at Wesleyan, I was fortunate to join Wesleyan's Center for the Humanities as a faculty fellow and Yale University's Information Society Project as a visiting fellow, and to participate in a symposium on innovation and the creative economy at the University of Texas at Austin. The feedback I gained from these experiences was invaluable. I thank the other participants for their willingness to entertain and nurture my nascent iconoclasm, and Ethan Kleinberg, Valerie Belair-Gagnon, and S. Craig Watkins for their organizational work and support. The foundations for this book were laid during my semester at the Center for the

Humanities. The article that grew out of this fellowship contains an earlier iteration of this book's argument; thanks to *New Media & Society* for publishing it. While at the symposium on innovation and the creative economy, I refined the argument presented in chapter 3. The paper I presented there contains an earlier iteration of that argument; thanks to S. Craig Watkins for editing the volume that emerged from the symposium, and to Routledge for publishing it.

Robyn Autry has been my constant companion at Wesleyan and throughout the writing process. She kept me on track, made finishing feel possible, and made me smile. She emboldened me to not tolerate bullshit, and has been a model for how to navigate a sometimes-hostile world fearlessly. I have learned more from being with her than from any text. Jonathan Cutler, too, has taught me more than he knows. He introduced me to the "antisocial thesis" and shook up my politics, and listened patiently as I tried to make sense of the texts examined in this book (and helped me to do just that when I struggled). His encouragement and confidence in me allowed me to stay true to my insight in the face of critical opposition. I couldn't ask for better colleagues or friends.

Despite my antisocial proclivities, no doubt exacerbated in the depths of writing, I somehow managed to hold on to my non-academic friends and queer family: Daniel Barrow, Katie Halper, Kate Levin, Cayden Lovejoy, Rebecca Ross, James Rubin, Kora Wilson, Avery Wolf Green, and August Wolf Lovejoy. I thank them for putting up with me, and for their kindness, brilliance, and humor. Also in the mix, Liz Montegary and Roy Pérez were excellent housemates.

Last but not least, I owe much of my political sensibility to my parents Isabel and Mark Goldberg—less from their well-versed socialism than simply from living life with them. They loved me and let me be, with a limp wrist and my head in the clouds. My sisters, Rita Strauss and Denise Ficker, were unrivaled models of rebellion, albeit unintentionally; I was a late bloomer in that department. My boyfriend, Ted Baab, has also been a model for me: of determination, humility, and grace, not to mention a devastating wit. It takes guts to be gentle and kind, and he is both, supporting me in ways both mundane and extraordinary, though I haven't always made it easy. Thinking by his side has been a privilege and a pleasure. With any luck he'll let me stay there a while longer.

# Introduction

According to scholars, journalists, and pundits, the institution of work is now undergoing three significant transformations at the hands of digital technology, all of which are essentially bad news for workers. First, the boundary between work and leisure is being eroded by the migration of leisure to online venues like Facebook and Instagram, rendering activities formerly outside the institution of work increasingly subject to economic exploitation. Second, jobs previously immune to automation are increasingly subject to automation, threatening not only these jobs but, insofar as these jobs represent the last vestige of human labor that cannot be automated, all jobs. Third, the so-called "sharing economy" preys on the unemployed and underemployed, offering low wages, few if any benefits, and no job security in exchange for renting underutilized assets like a spare room in one's home (Airbnb), or for doing odd jobs or chauffeuring in one's spare time (TaskRabbit, Uber, Lyft), the coordination of which is facilitated through Internet and mobile applications.[1] Considered together, these transformations would seem to suggest that it is not only workers that are now under attack, but the very institution of work itself.

This book is not so much about these imagined transformations—whether to confirm or deny their empirical validity—as it is about the hand-wringing that both accompanies and, I will argue, structures these imagined transformations. Through an in-depth examination of concerns about "playbor" (as the collapse of work and play has been called), automation, and the sharing economy, the book makes a series of related arguments: (1) that these concerns are an expression of anxiety, and that understanding them as such helps to uncover deeper, underlying concerns that have gone unstated and thus unexamined and unquestioned; (2) that these deeper, underlying concerns are not about the material well-being of workers (as they appear to be on the surface), but rather about the erosion of particular forms of relationality valued by critics—

collective, communal, responsible, accountable, sacrificial, and so on—in a word, social; and (3) that framing scholarly and popular concerns surrounding these transformations as expressions of anxiety helps to illuminate how they are part and parcel of a normative project that aims to produce the very social subjects who are supposedly endangered by these transformations. In other words, the book aims not only to identify this normative project as such, but to explore how this project is furthered through an anxious response to the loosening of social bonds. It argues that concerns about playbor, automation, and the sharing economy serve both as a smokescreen for valuing the social, and as a vehicle for this valuing and its normative ends.

Following the affective turn in social theory, the book thus explores how a particular "feeling"—anxiety—can function as a political strategy or technique, rather than simply an internal psychological state.[2] Anxiety functions in this way by initiating specific interventions in a world perceived as worrisome, interventions that aim to resolve the underlying cause of anxiety. In elaborating this argument, the book draws from Sianne Ngai's and Sara Ahmed's work on emotion and anxiety, which usefully theorizes the epistemic (rather than ontological or practical) character of these interventions.[3] To take one example, chapter 3 argues that anxieties about automation and job loss do not primarily aim to curtail automation or to ameliorate its negative economic effects—what seem to be the apparent goals of the texts examined—so much as they aim to assert the value of collective forms of governance.

The book is also animated by a particular set of feelings—my own: the shame of desiring and taking pleasure in the antisocial or asocial (and in failing to identify with the social), anger at the imposition of valued forms of relationality, particularly through norms, and by the disavowal of this imposition, but also—as one moves away from these valued forms—a certain indifference toward norms that work, in part, through the mobilization of shame.[4] Because values are social in nature, the value of the social itself can only be posited from within the social.

While working on this book, these feelings found a home in what has contentiously been called the "antisocial thesis" in queer theory, a line of thought that identifies the antisocial character of queerness as its most radical political utility, without simply transforming the antisocial into a positive value. (Again, the "social" of antisocial is meant to describe valued forms of relationality; in this framework, one can be relational without being social, as in various forms of anonymous, casual sex, for example.[5]) The antisocial thesis is most often associated with queer theorist Leo Bersani, particularly his 1987 essay, "Is the Rectum a Grave?" and 1995 book, *Homos*, as well as Lee Edelman's 2004 book, *No Future: Queer Theory and the Death Drive*. Written across two decades in which respectability politics became increasingly popular—first as a response to the stigmatization of "gay sex" during the HIV/AIDS crisis in the 1980s and 1990s, and then as mainstream LGBT advocacy groups prioritized the legalization of same-sex marriage and inclusion in the military in the 1990s and 2000s—these texts not only blatantly refused the directive that queers become respectable, but asserted the political importance of a queer inability to become respectable, to follow norms, to assert values, to participate in community—in short, to be properly social subjects. If this has been an uneasy prospect for many queer scholars on the Left, even the radical Left with its panoply of social commitments, it has nonetheless proven to be an irresistible object of thought and debate, in part because it appeals to a longstanding queer interest in and desire for freedom from governance.

The antisocial thesis, particularly as formulated by Bersani, not only helped to legitimate the feelings, desires, and pleasures that animate the argument of this book; it also provided a theoretical foundation for identifying, examining, and critiquing the imposition of the social. The book is thus grounded by the contention that if to be queer is not simply to be abnormal, but to be opposed to the normative as a method of social control, then this opposition ought to include the normativity of sociality, particularly insofar as normativity is a function of the social.[6] It is also grounded by the contention that an antisocial critique is made more urgent by what seems to be a widespread disinclination to identify the imposition of the social as such, and by the notion that this identification could engender different and desirable queer politics.

In sitting with these feelings, and the thoughts and arguments they inspired, I have sometimes thought of this book, with tongue somewhat in cheek, as a kind of "coming out," in that it questions the valuing of community and other forms of collective relationality at what seems like a particularly inopportune moment, or at least an unpopular one, to do so.[7] With increasing wealth inequality and the persistence of social injustices and inequities, environmental catastrophe looming, and much of the radical Left with neoliberalism in the crosshairs, to be critical of the social is, for some, to support (if inadvertently) the wrong side in its many guises—capital, capitalists, the market, multinational corporations, and so on, who are seen as sharing this disregard for the social—and also to be insufficiently distressed by those social, political, and environmental issues against which forms of collective relationality are often mobilized. One might think here of the long history of anti-Black violence in the United States and, in response to this violence, the Black Lives Matter movement—to take just one critical, contemporary example. As this book goes to press, the Left is up against Donald Trump's presidency, which has provided yet another rationale for the strengthening of social ties as a necessary condition for resisting Trump's heteropatriarchal, classist, white supremacist discourse and agenda.[8] In short, to be critical of the social is to be seen as aligned with or even to further oppression, subjugation, and exploitation, as if valued forms of relationality were not also imposed, in part through their very valuing.

Connected to the notion that criticism of the social is insufficiently attuned to injustice and inequity, and perhaps most damning of all objections to the antisocial, is the idea that it is a privilege few can afford—and presumably not the kind of privilege that ought to be shared, but rather destroyed. Perhaps more to the point, even the appeal of the antisocial has become suspect as a marker of privilege, like a high-pitched whistle that can be heard only by affluent, gay, cisgendered, white men, despite a number of affinities between the antisocial thesis and other strands of queer scholarship.[9] This characterization of the antisocial thesis as elitist and racially obtuse has been advanced most forcefully by José Esteban Muñoz in his 2009 book, *Cruising Utopia: The Then and There of Queer Futurity*, as well at a 2005 MLA panel at which Muñoz, Edelman, Robert L. Caserio, Judith Jack Halberstam, and Tim Dean sparred over the antisocial thesis.[10] While Muñoz was sympathetic toward much of Edelman's argument, despite their obvious political differences, Muñoz's critique of the antisocial thesis as not just insensitive to other tropes of difference (other than sexuality, that is) but as hostile to intersectionality would effectively squash whatever intellectual good will the antisocial thesis had previously entertained, especially with the contemporaneous formulation of queer of color critique.[11] It seems likely that it is for this reason that the antisocial thesis is a doctrine almost entirely devoid of subscribers, as Tim Dean has suggested.[12]

For those excluded from privilege, and particularly for those facing "social death," access to the social has often been thought (by the Left) to be empowering, as are the subject positions or identities established through this access.[13] If those of us who wield various kinds of privilege cannot simply divest ourselves of our privilege—so the argument goes— the least we can do is act as allies to those without privilege, engaging in a process of alignment with the disenfranchised other that works to bring this other into the social and, in so doing, to absolve partially the privileged subject of the ethical burden of its privilege, bringing it back into proximity with the good. Certainly many scholars have taken this route. In this solemn and righteous academic climate, has it become im- possible to make a case for what we might think of, in contrast to social death, as social suicide—a voluntary opting-out of the social? Has it be- come impossible to see the social as a prison rather than as a privilege?[14] Easy for me to say, so the critique goes.

This book is premised on the notion that it is possible to make such a case via the antisocial thesis and, furthermore, that doing so could prove useful in identifying and ultimately transforming relations of power.[15] For example, the attachment to work as an institution through which the social is established and maintained helps to support that institu- tion; to leave behind this attachment—to detach—might draw us closer to a world without work, a political vision that has generated interest in recent years.[16] The book thus aims not only to expose and critique the unidentified and underexamined conservatism of the Left as it is expressed through the anxiety surrounding the transformation of work at the hands of digital technology, but in so doing also to demonstrate the "usefulness" of the antisocial thesis, a doctrine seemingly without subscribers.[17]

That said, following the antisocial thesis, the book does not aim to offer a positive program of political resistance, tangled as such visions tend to be with the very same social forms that are the antisocial thesis's object of critique. Rather than a positive political program, what the an- tisocial thesis offers might be better described as a position of critique.[18] For example, the book does not embrace as revolutionary or liberatory those purported transformations to the institution of work that trigger critics' anxiety. Indeed, the book neither confirms nor refutes that these transformations are actually occurring. Rather, it seeks to show how critics' understanding of and anxious response to these transformations serve a normative project. In this way, the intervention the book aims to make might best be described as a "calling out," a gesture that I hope will make space for the antisocial, or rather—insofar as the "anti" remains tied to that which it is against—for something like indifference to the social. If there is refusal in this gesture, it is not a grand political refusal, with the positioning, posturing, and organizing this would entail; instead, it is a refusal to transform into that which the lines of criticism examined below would have us become.

* * *

In order to redescribe concerns about playbor, automation, and the sharing economy as expressions of anxiety with normative ends, the first chapter of the book ("Anxiety and the Antisocial") elaborates a novel theoretical framework, drawing from the antisocial thesis, particularly as formulated by Bersani, as well as recent work on affect and emotion by Sianne Ngai and Sara Ahmed. Weaving together these theoretical strands, the chapter proposes that anxiety is not simply an individual psychological disposition, but can also be ascribed to modes of thought. The chapter then argues that anxiety, as a discursive affect, functions as a "straightening device," policing antisocial subjects (or non-subjects) and calling them back to valued forms of relationality.[19]

The following three chapters explore the anxiety surrounding technologically enabled transformations of the institution of work through three case studies. In each of these cases, the book gathers together a series of academic and popular texts—often already in conversation with one another—to provide a foundation for analysis and interpretation. While these texts are not homogenous in approach, perspective, or argument, they are nonetheless often structured by what the book argues is a uniform valuing of the social, the presence of which can be uncovered by examining the anxieties that surface in each case. The second chapter ("Playing") examines anxieties surrounding the transformation of leisure into work—particularly forms of online leisure like social networking that are productive of economic value—as these anxieties have been expressed in a series of contemporary academic texts. In this body of work, critics contend that the leisurely façade of life online masks the economic exploitation of users, on whose backs companies like Google and Facebook have amassed tremendous wealth. The second chapter also considers the related critique of what I term "leisure-at-work," as when "creative class" or "no-collar" employees engage in various kinds of leisure—for example, playing games, getting massages, and goofing off—while at the office and with the consent and blessing of their employers. The third chapter ("Automating") examines anxieties surrounding the automation of human labor, as expressed in both popular writing and academic scholarship. The texts considered in this chapter argue that advances in robotics and machine learning are driving a new wave of automation, threatening forms of human labor formerly thought to be immune to automation, perhaps even threatening all forms of human labor. While such fears continually arise and are disproved by history, critics contend that this time will be different. The fourth and final chapter ("Sharing") examines anxieties surrounding the sharing economy, as expressed in popular and academic arguments that sharing-economy jobs offer workers a raw deal—lower salaries, fewer benefits, and little job security—and that workers were essentially forced to take these jobs in the wake of the Great Recession—all of which has been masked by a veneer of communitarianism in sharing-economy rhetoric.

On the surface, the concerns expressed in these three lines of argument might seem to have little to do with the dismantling of the social. Rather, they would seem to be about the state of work and the lives and wellbeing of workers, written in an effort to expose and denounce exploitation and promote the growth of satisfying, sustaining jobs. Why, then, imagine that something else is going on beneath the surface of these arguments and their stated concerns? The impulse to interpret these concerns, rather than accepting them at face value, is motivated by inconsistencies internal to each argument. The second chapter identifies inconsistencies in the redescription of leisure as labor: only certain forms of value-producing leisure register for critics as labor, though by critics' own definition of labor, many other forms of leisure ought to be included, for example forms of leisure paid for with money rather than time. The third chapter identifies inconsistencies in concerns about unemployment and underemployment: in the scholarly and popular texts examined, only automation is identified as a cause for unemployment and underemployment, though if this were truly critics' concern, many other causes ought to register as significant—the outsourcing of jobs overseas, falling rates of consumer spending, and so on. Finally, the fourth chapter identifies inconsistencies in concerns about the sharing economy: scrutiny is directed toward sharing-economy businesses, but not their traditional market counterparts, despite fundamental similarities. In addition, critics argue not only that sharing-economy laborers are economically exploited, but also that the kind of services sold in the sharing economy ought to be offered for free for the good of the community; workers are both not compensated enough and should not be compensated at all.

These inconsistencies open up a central, guiding question of the book: If it is not (or not simply) the securing of workers' livelihoods and well-being that primarily animates critics, what are they really concerned about and to what end? When a feared object is revealed to be a phantasm, we can ask: What is really feared? As Barry Glassner suggests, "Fear, like desire, tells us very little about its object."[20] We might say the same of anxiety, especially insofar as anxiety characteristically attaches to many and varied objects.

Through textual analysis and interpretation in the mode of "symptomatic reading," the book identifies and analyzes a structuring, underlying anxiety in each of these three cases. The second chapter argues that concerns about the exploitation and alienation of "playbor" and workers-at-play are motivated by an underlying discomfort with forms of leisure and pleasure understood as self-indulgent and irresponsible. The third chapter argues that concerns about the extinction of labor at the hands of technology conceal an underlying attachment to governance, whether the governance of the state (which becomes necessary in order to ameliorate the effects of automation) or collective governance (against which our robot overlords serve as a discursive foil). Finally, the fourth chapter argues that concerns about the increasing precariousness of labor in the sharing economy are rooted in a rejection of the market and money as inimical to valued social bonds, insofar as money engages actors in antagonistic, self-interested exchange rather than in the supposedly communal, altruistic relations of gift-giving. Considering these three cases together, what emerges is a picture of valued forms of relationality—collective, responsible, sacrificial, accountable, and self-governing—under threat by technologically enabled transformations to the institution of work.

The book's epilogue considers one final and overarching anxiety that extends beyond the institution of work: the anxiety that social and political-economic life have been rendered less material by contemporary forms of media and digital technology, and that this immateriality represents a threat to social and political-economic stability and well-being. To evidence this anxiety as it has been expressed across different registers, the epilogue gathers together an "archive" that includes Spike Jonze's 2013 film *Her*, speeches given by Barack Obama, and contemporary advertising campaigns. Considering these sources together, and ruminating on Marx and Engels's famous statement that "all that is solid melts into air," the epilogue proposes that the nostalgia for materiality given voice in these texts might be better understood as a desire for the social bonds that have diminished in our apparently immaterial world; the material, in other words, serves as a discursive proxy for the social.

The book argues that in all of these cases, the anxiety surrounding the loss of social bonds does not simply express an attachment to these bonds, but is part and parcel of an attempt to establish or restore these bonds and to eliminate their constitutive other: the self-involved, non-committal, promiscuous, irresponsible, pleasure-seeking, antisocial, non-subject—a persona non grata par excellence. In each of the three cases examined, the expressed concerns—about playbor, automation, and the sharing economy—do not simply target (respectively) the exploitation of leisure, the elimination of jobs, and the precariousness of labor; they also and perhaps more fundamentally target irresponsible pleasure, ungovernable anomie, and the dissolution of social bonds, in an effort to call readers back to the social.

1

# Anxiety and the Antisocial

According to critics, the years following the Great Recession of 2007–2009 were not good for most workers in the United States. A prototypical worker during this period might be described as follows: lost his or her job shortly after the housing bubble burst and the market collapsed; subsequently unable to find steady, full-time employment, in part because "good jobs" are increasingly automated; as a result, forced to drive for Uber, or rent out his or her apartment/spare room/sofa through Airbnb, or work odd jobs through TaskRabbit to make ends meet, though some of these "jobs" will also soon be automated (it is only a matter of time until Uber cars are driverless); and finally, driven by the resulting stress or restlessness to distraction online, the economic value of which is siphoned away by the very same industry that is both automating good jobs and replacing them with worse jobs or "gigs" (as critics sometimes refer to them).[1]

As detailed in the chapters that follow, the handwringing that accompanies these purported transformations is considerable. How should we make sense of these concerns: that our leisure online is being transformed into work and is exploited as such, that the Internet has made possible a so-called sharing economy that produces only precarious jobs, and that all jobs—precarious or not—may soon be automated? Should these concerns be engaged empirically, tested against evidence that might confirm or refute the observed transformations? I take a different approach here, inspired by what I identify as inconsistencies internal to these concerns/arguments, and informed by critical analyses of knowledge production, which highlight the ways that empirical accounts often mask their underlying, structuring motivations and ends.

To take one example of this kind of analysis of knowledge production, in his critique of ethnographic methodology, Vincent Crapanzano identifies a central paradox constitutive of ethnographic research: ethnographers often confess the provisional nature of their interpretations, but do not acknowledge the provisional nature of their observations—what Crapanzano calls "presentations"—though these observations acquire their legitimacy precisely through interpretation.[2] As Crapanzano writes, "Embedded in interpretation, [the ethnographer's] presentations limit reinterpretation."[3] Following Crapanzano, the question that guides the inquiry of this book is not whether or to what extent the institution of work is being transformed or dissolved, but rather what lies beneath these "presentations," with their originary, embedded interpretations?

Eve Kosofsky Sedgwick's notion of "paranoid reading" also suggests a mode of analysis that circumvents questions about the truth value of empirical claims and focuses instead on what such claims do, their performative effects.[4] For Sedgwick, the practice of paranoid reading is characterized not simply by its particular negative affect, but by the way it disavows this affect and masquerades as "the very stuff of truth."[5] In this mold, the aim of the book is not to arrive at a more definitive empirical accounting of the observed transformations, but rather to leave aside the issue of veracity in order to open up questions about the motivations and ends of the texts examined. To put it another way, for my purposes here, the examined texts are less interesting as a window into the world, than they are as a window into a particular way of seeing the world and, ultimately, of intervening in it.

This way of seeing has two prominent and related characteristics: a particular feeling or affect and particular attachments and investments. Raymond Williams's concept of the "structure of feeling" is useful here; it describes "a pattern of impulses, restraints, tones"—feelings or affects rather than thoughts—shared between texts that do not appear to be otherwise connected to each other.[6] According to Williams, rather than simply looking for shared ideas, it is instructive to look across texts for shared feeling or affect, and for the attachments and investments that these serve.

The texts examined in the chapters that follow are structured by a series of topically related concerns—that leisure is exploited, that jobs will disappear, and that the jobs that remain are degrading in quality—as well as by what I will argue is a shared feeling or affect of anxiety. Why anxiety? All concern is, in a sense, anxious in nature. To be concerned is to be worried. But, as the first half of this chapter explores, anxiety as a feeling or affect can be distinguished in part by its deceptive nature, that is, by the way that it obfuscates its true object. To put it another way, the actual object of concern is not the expressed object of concern. This means that if, as I have suggested, empirical texts often mask their underlying, structuring motivations and ends, then these motivations and ends are doubly masked in anxious texts. So while these texts appear to be concerned about economic exploitation, job loss, and "precarity"—a term that describes the fragmentation and discontinuity of work relations and experiences in contemporary capitalism—I will argue—via contemporary theorizations of anxiety—that they are more precisely concerned about the dissolution of social bonds. Just as the expressed concerns examined here (about exploitation, job loss, precarity) are deceiving, so too are the attachments these objects would suggest. It is not, or not simply an investment in workers' livelihoods and well-being that lies at the center of this anxiety, but rather, I will argue, an attachment to responsible forms of relationality, and all that these forms entail: sacrifice, discipline, self-governance, accountability, and so on.

To be certain, this is not to say that there has not been anything to worry about in the wake of the Great Recession.[7] When workers lose their jobs—as characteristically happens during an economic downturn—and have neither wealth nor other resources to fall back on, nor job prospects, but rather expenses to cover, debts to pay, and/or friends/family to support, there is a lot to worry about. Furthermore, as the effects of an economic downturn are unevenly distributed—for example, by race, sex, gender, sexuality, and class—so too should we expect worry to be unevenly distributed; as some people are more economically vulnerable than others, so too do they have more reason to worry. But this worry might be better characterized as fear rather than anxiety (as I define it below). In any case, it might be distinguished from the anxieties examined in this book by the consistent identification of potential threats—that is, when all potential threats are registered as such (in relation to the object of worry)—in contrast to the inconsistencies detailed in the Introduction and described in more detail in the chapters that follow.

Nor is it to say that there are not legitimate reasons to oppose the corporations, social practices, or government policies that motivate some critics; the book certainly does not aim to offer a defense of playbor, automation, or the sharing economy. For that matter, neither does it contain an exhaustive account of the illegitimate or deceptive reasons one might oppose these. My scope here includes neither an analysis of these other reasons, nor an account of the empirical implications of automation, playbor, or the sharing economy. Rather, my argument here addresses a specific set of concerns—selected for their prominence during the years following the Great Recession—that I propose be understood using the concept of anxiety for reasons articulated below.[8]

Drawing from work on anxiety by Sara Ahmed and Sianne Ngai, the chapter theorizes anxiety as a particular way of seeing/feeling but also of intervening, of attempting to resolve the underlying cause of anxiety by producing a subject invested in or attached to desired objects in particular ways. In terms of the cases analyzed in this book, expressed concerns about labor mask an underlying anxiety that does not only or primarily aim to improve conditions of labor, but rather to establish or secure valued forms of relationality by soliciting readers to identify as responsible subjects. Anxiety is thus not only the feeling or affect of "digital dystopianism"—that is, the perspective that understands digital technologies as imperiling social, political, economic, and psychological well-being. It is also the means through which this perspective attempts to secure and further that to which it is attached or in which it is invested: in this case the properly social, responsible subject. Understanding dystopian concerns through the framework of anxiety, in turn, draws into focus dystopianism's attachments and investments and, thereby, its underlying aims and motivations.

To be clear, I do not meant to suggest that the authors of these texts are themselves anxious, and that the texts they write express this anxiety, though this is certainly possible. Rather, following Williams, I see the texts themselves as being structured by the feeling or affect of anxiety. As Sedgwick suggests, the (paranoid) affect of a text describes a particular position or practice, rather than the psychological state of its author. Similarly, Patricia Clough has argued that it is both possible and useful to conceptualize thought outside the intentions of authors. As Clough writes, "Thought is not given by individual thinkers so much as it is given to them as they are drawn to the future by it."[9] This is to attribute a kind of agency to thought, rather than to understand it as simply the product of individual authors; in this conceptualization, thought takes shape "outside subjectivity, even outside human intersubjectivity," and exhibits "its own movement, intensities, and affects."[10] In conceptualizing thought outside human intersubjectivity, Clough ascribes a kind of agency to thought that exceeds not only the intentions of authors but also the meanings created by readers, who might be similarly reconceptualized as drawn to the future by thought.

It is important to note here that Sedgwick's account of paranoid reading is meant not as a positive contribution to the tradition (in literary criticism) of "symptomatic reading," in which primarily fictional texts are hermeneutically mined for hidden or latent truths and for the ideologies served by these, but rather as a critique of this tradition. The paranoid, for Sedgwick, is not the analyzed text but the analyzing text; it is literary criticism that is under scrutiny here. Sedgwick's account, then, would seem to implicate precisely the project of this book, which offers a nonobvious interpretation of concerns about playbor, automation, and the sharing economy. Furthermore, Sedgwick's critique is but one iteration of what appears to be a contemporary trend away from symptomatic reading as a critical methodology, from Slavoj Žižek's proclamation in *The Sublime Object of Ideology* that symptomatic reading is impossible, to a 2009 special issue of *Representations* ("The Way We Read Now") devoted to rethinking and offering alternatives to symptomatic reading.[11]

In part, the shift away from symptomatic reading has been explained as an effect of the increasing salience of outright expressions of power like the torture of prisoners at Abu Ghraib and the United States government's non-response to Hurricane Katrina; when the exercise of power is so blatant, the argument goes, reading beneath or beyond the surface is unnecessary.[12] As Sedgwick asks, "Why bother exposing the ruses of power in a country where, at any given moment, 40 percent of young black men are enmeshed in the penal system?"[13] Indeed, in their introduction to "The Way We Read Now," Stephen Best and Sharon Marcus suggest that the very project of politically radical literary criticism was compromised by the geopolitical landscape of the early twenty-first century, as if literary criticism were suddenly rendered obsolete by the Bush regime, but literary critics still had to show up for work anyway. Rather than unmasking the conservative truths that lurk beneath the deceptive surfaces of the text, some critics now prefer simply to describe the text, appreciating the complexity of its surface without either celebrating or denouncing its supposedly hidden politics.

If literary criticism once provided theoretical tools that could fundamentally draw into question the empirical aspirations of the social sciences, it now flirts with precisely this empiricism.[14] While this shift seems to make criticism less necessary or important, it also establishes a less paternalistic and more humble role for the critic, who is no longer required to decode the secrets of the text, as well as relieving literary criticism of any kind of utilitarian function, most notably the pursuit of freedom, albeit from a resigned perspective. There is thus a curious tension in the shift away from symptomatic reading, which seems to offer both a capitulation to the "real"—one manifestation of the Left's retreat from cultural radicalism after the Culture Wars, the so-called Sokal Affair, and 9/11—and an argument for idleness, a vestigial remnant of that same radicalism, perhaps. While I am sympathetic to the latter argument (for reasons that will soon become clear), the former argument warrants some scrutiny.

The claim that power is now out in the open imagines power as a surface that can be easily read, written on the bodies of the tortured, the incarcerated, the exploited, and the abandoned. It is unclear whether for critics such forms of power are now dominant (having displaced earlier forms) or are simply more important than the hidden, ideological forms that had previously concerned them. In either case, there is a retreat from the symbolic register—in which meaning is hidden and needs to be drawn out—which makes the contemporary interest in surfaces and surface reading seem rather disingenuous. In other words, if symptomatic reading has fallen out of favor, it is not because surface and depth have collapsed in such a way that would render symptomatic reading pointless, and surface reading a worthwhile exercise, but rather because surfaces no longer seem to matter in a world in which the real has become obvious.

It is precisely the Left's claims of access to the real that make necessary something like symptomatic reading, not because these claims are false—again, my interest here is not in verifying or falsifying empirical claims—but because they are vehicles for an unidentified and unexamined normative project. While these claims have relieved literary critics of the burden of heroically exposing the truth, making possible a less utilitarian if somewhat guilty politics within literary criticism, they also provide a warrant for strengthened forms of communal, collective relations, which become necessary to oppose the status quo; the heroic critic has been replaced by the heroic community. This power play by the (radical?) Left seems remarkably under-examined. Indeed, it rarely if ever seems to register as power at all, as when Best and Marcus suggest that the notion of veiled domination is nostalgic. And while it is rather ironic that the insistence on the visibility of particular forms of power can itself be used to mask other forms of power, this should not come as a surprise, insofar as the Left generally aspires to inclusion; its preferred mode of power is not violent but benevolent, seemingly motivated by a concern for the well-being of those whose normative acquiescence it will demand in the last instance.

In order to account for disavowed normative modes of power, the second half of the chapter turns to queer theory, with its longstanding interest in (if not commitment to) antinormativity. Responding to contemporary debates about the status of antinormativity in queer theory, the chapter argues for the continued relevance and utility of an antinormative perspective or politics, particularly as expressed via the so-called "antisocial thesis," a much maligned vein of thought that draws into question the normativity of the social itself and theorizes valued forms of relationality as a mode of oppression or subjugation. Drawing in particular from work by Leo Bersani, the chapter proposes that the Left's call to the social—including by scholars working within queer theory—is motivated not by an altruistic or empathetic care for the other, but rather by a desire to annihilate the other's difference through incorporation or assimilation. The chapter offers "indifference to difference"—to borrow phrasing from Madhavi Menon—as an alternative to the social, which allows for a kind of relationality without this endemic violence.[15] The chapter thus brings together the antisocial thesis (as well as queer theory's general critique of norms as mechanisms of power) with theorizations of anxiety in affect theory in order to argue that the anxiety that surrounds the dissolution of the social functions as a kind of "straightening device" that aims to bring the asocial or antisocial non-subject back into line.

## Ages of Anxiety

The notion that we live in an age of anxiety is not new. As Daniel Smith writes in the opening essay for the *New York Times'* anxiety column,

"'The age of anxiety' has been ubiquitous for more than six decades now," beginning (famously) with W. H. Auden's celebrated and intractable 1948 poem of the same name.[16] While some of this might simply be hype—scholars, authors, and artists seem to have a thing for the naming of ages (for example, Thomas Paine's *The Age of Reason*, Edith Wharton's *The Age of Innocence*, and Walter Benjamin's "The Work of Art in the Age of Mechanical Reproduction")—even a perfunctory stroll through culture suggests the salience of anxiety as a dominant way of understanding and responding to the modern and contemporary world.

In the years following the financial crisis of 2007–2008, anxiety became a particularly popular catch-all to describe various kinds of contemporary malaise, as illustrated, for example by the *New York Times* anxiety column, which was published regularly between January 2012 and July 2013 (and then revived in February 2015), as well as a number of books, from memoirs like Daniel Smith's *Monkey Mind: A Memoir of Anxiety*, Patricia Pearson's *A Brief History of Anxiety (Yours and Mine)*, and Scott Stossel's *My Age of Anxiety: Fear, Hope, Dread, and the Search for Peace of Mind* to academic and popular texts in psychology like Allan V. Horwitz's *Anxiety: A Short History* and Jeffrey P. Kahn's *Angst: Origins of Anxiety and Depression*. The list grows longer if one includes anxiety's cousin, depression. An argument could easily be made for including Ann Cvetkovich's *Depression: A Public Feeling*, as well as Sianne Ngai's *Ugly Feelings* and Sara Ahmed's *Collective Politics of Emotions*, the latter two of which inform the theoretical perspective elaborated here. The publication of these texts bespeaks the contemporary popularity of anxiety as both a diagnosis and object of interest.

Much of the recent interest in anxiety has been clinical and pragmatic, focused on cause and cure or treatment. In these accounts anxiety is understood as a psychological condition, affliction, pathology, or disorder characterized by a concatenation of symptoms: difficulty concentrating, feelings of dread, apprehension, and powerlessness, nervous habits, elevated heart rate, rapid breathing, insomnia, and so on. Contemporary accounts of anxiety are grounded in and bolstered by statistics that construct anxiety as a mental health epidemic. Smith, for example, cites a figure published by the National Institute of Mental Health: 18 percent of the adult population in the United States (about 40 million people) suffer from anxiety, outpacing all other mental disorders by a wide margin.[17] In *Angst*, Kahn groups together anxiety and depression, arriving at a slightly larger figure: 20 percent of (all) Americans, or 60 million people.[18] These statistics, and the epidemic they construct, shape contemporary notions of our current age of anxiety as one in which symptoms of anxiety become legible as such, and in which the experience of anxiety as an assemblage of these symptoms becomes measurable at the level of the population.

Some scholars have responded critically to the medicalization and "epidemicization" of anxiety, highlighting how the diagnosis and treatment of anxiety is geared toward normalization and regulation at both the individual and population levels, as Jackie Orr argues.[19] In this sense, the medicalization of anxiety has a disciplinary function (in the Foucauldian sense of the term): the management of anxiety becomes an occasion for regulating bodies. Anxiety becomes something to contain, though perhaps not eliminate insofar as it is economically productive. Similarly, in other disciplinary fields in which anxiety has been examined, including philosophy, theology, and sociology, anxiety remains a problem in search of a solution. It is something that needs to be resolved, in order to return the anxious subject to affective equilibrium.

My approach to anxiety differs from these accounts in two important ways. First, I do not take up anxiety as a psychological condition, that is, as an experience of symptoms that are suffered by individual people, but rather as an affect that suffuses and structures particular modes of thought. Second, I understand anxiety as productive or active, making desired interventions in the world, rather than as a psychological obstacle or impairment that needs to be resolved, or whose management is productive in terms of disciplining/controlling bodies and populations. In this way, I suggest, anxiety can also be understood as a technology of power, producing particular kinds of subjects. This is not to dispute that psychological "dis-ease" is manufactured so that, for example, pharmaceutical companies can sell more SSRIs, as Orr argues, but rather to supplement such accounts with a notion of anxiety that can account for the ways that it is desirable as a method of subject formation. To put it another way, anxiety is not simply a psychological problem to be managed or eliminated, but a desirable affect to be produced.

This conceptualization of anxiety is informed by contemporary work on affect, which has suggested that feelings and emotional states are not simply private psychological dispositions, but rather "mediate the relationship between the psychic and the social, and between the individual and collective," as Sara Ahmed has written.[20] Drawing from this body of work, and in particular from the scholarship of Ahmed and Sianne Ngai, I aim to conceptualize the ways that anxieties about the transformation of work are productive, working to establish or reestablish subject boundaries that are thought to have been breached and on which valued forms of relationality depend.

Rei Terada and Sara Ahmed both point out that the word emotion comes from the Latin *e-movere*, meaning "to move out" or "to migrate."[21] But Ahmed rejects the notion that feelings form within selves and then move outward, a notion popularized by psychological discourse. She also contests the sociological notion that feelings can form outside the self, as in collectives or crowds, and then penetrate the self, moving inward. In contrast to these two models, Ahmed suggests that "emotions work to create the very distinction between the inside and the outside, and that this separation takes place through the very movement engendered by responding to others and objects."[22] In this way, Ahmed denaturalizes the boundary between inside and outside the self, often taken as obvious. Along these lines, she also notes that emotions can produce the sense that one's own self is discontinuous, as when menstrual cramps feel like they come from one's own body and simultaneously like they are foreign.[23]

Ahmed's argument here offers an alternative to the commonplace notion that emotions felt within the self are caused by objects and others outside the self, instead proposing that emotions work to erect the very boundaries between these entities.[24] She elaborates: "It is through emotions, or how we respond to objects and others, that surfaces or boundaries are made: the 'I' and the 'we' are shaped by, and even take the shape of, contact with others."[25] This presents an apparent paradox: How can a self respond to others and objects if its borders have not yet been established, that is, if its inside has not been distinguished from its outside? To put it another way, how can contact occur between two entities if they have not yet been bounded as entities? The word "contact" makes sense only in the context of bounded entities. Entities that have not been bounded are amorphous or blurry, and blurs do not exhibit points of contact; they do not touch, they blend or bleed.

At the center of this paradox is the ontology of the boundary. Do boundaries between selves, others, and objects exist in the absence of the feelings that register them? If they do, then feelings simply report and confirm the presence of boundaries. If they do not, then feelings actively produce boundaries. Ahmed seems to indicate not only that "emotions work to create the very distinction between the inside and the outside" but also that emotions and sensations (which work together and inseparably in Ahmed's account) do not produce the surface of the body.[26] As she writes in her discussion of pain, "It is not that pain causes the forming of the surface. Such a reading would ontologise pain (and indeed sensation more broadly) as that which 'drives' being itself."[27]

How is it possible that emotions "work to create" surfaces or borders while not "causing" them to form? In order to resolve or perhaps sidestep this paradox, we might bypass the language of production and causality altogether. In *Meeting the Universe Halfway: Quantum Physics and the Entanglement of Matter and Meaning*, Karen Barad offers a dramatically reconceptualized notion of matter that does not presume that matter preexists measurement or that it is produced through measurement.[28] Rather, Barad proposes that matter assumes no determinate form prior to measurement; it is through processes of measurement (taken broadly to include things like observation and description) that matter assumes a particular form. Following this reconceptualization of matter, Barad rejects the notion of interaction, which presumes the preexistence of determinate matter that might be set into relation, and replaces it with the notion of "intra-action" to account for the way that measurement determines matter's form or, in other words, the way that measure informs matter.

Following Barad, emotions, affects, and sensations might be characterized as a kind of "measuring agency" that works to determine the boundaries of intra-acting selves, others, and objects. From this perspective, emotions do not inhere in a pre-constituted self, but rather in indeterminate matter—matter that has the capacity to register local events, shifts, or movements as contact. When activated through emotions, this capacity determines a boundary between self, other, and object. In short, emotions do not create borders or surfaces, or cause borders or surfaces to form; rather, they determine or "perform" them. We might thus amend Ahmed's argument: it is not contact with objects or others that generates emotion, but rather an indeterminate event measured (that is, determined) through emotion, affect, and sensation as contact between self and other or object.

Like other emotions or affects, anxiety can be theorized as a method of determining boundaries between self, other, and object, though in a particular way, with particular effects and ends. Ahmed's and Sianne Ngai's theorizations of anxiety help to clarify these effects and ends. Following Ernst Bloch, Ngai argues that anxiety has a distinctly temporal character; it is an "expectant emotion."[29] Like fear, anxiety does not focus on the present, but on the future. Its "drive-object" does not lie in the "already available world" but rather in a future configuration of the world and of the self. As Gordon D. Marino notes (in his explication of Kierkegaard's conceptualization of anxiety), it is the open-endedness of the future that makes it a repository for our anxiety.[30]

While fear and anxiety share several characteristics in common, in psychological discourse anxiety is typically defined in distinction to fear. Unlike fear, which is thought to have an object insofar as we are afraid of particular things—the dark, heights, spiders—anxiety has no predefined object. However, as Ahmed explains, anxiety attaches to objects in practice: "In anxiety, one's thoughts often move quickly between different objects, a movement which works to intensify the sense of anxiety. One thinks of more and more 'things' to be anxious about."[31] Rather than conceptualizing anxiety as fear without an object, Ahmed therefore draws a different distinction: "Anxiety becomes an approach to objects rather than, as with fear, being produced by an object's approach."[32]

In this sense, anxiety has a distinctly spatial character as well. It is oriented not simply toward the future but toward others and objects and, crucially, the self in relation to these. In psychological discourse, anxious subjects characteristically misattribute the source of their anxiety, displacing it from the self and projecting it onto others and objects. Projection, in this usage, describes "a quality or feeling the subject refuses to recognize in himself and attempts to locate in another person or thing."[33] Importantly, this is not to imply that subjects first become anxious and then project their anxieties outward. As Ngai argues, projection does not displace a preformed anxious condition, but rather is "the means by which the affect assumes its particular form."[34] This particular form is marked by the subject's loss of fixed position, that is to say the subject's inability to determine the boundaries between inside and outside, self and other. For this reason, it makes little sense to speak of an anxious subject preceding the act of projection (an act that characterizes anxiety); such a subject cannot exist since it is the act of projection that determines its boundaries. As Ngai argues, the act of projection reinforces the distinction between center and periphery, here and there, securing "a strategic sort of distance for the knowledge-seeking subject . . . even in the absence of the fixed positions from which nearness and farness are ordinarily established or gauged."[35] Ngai draws this insight, in part, from Freud, who initially understood anxiety as an internally constituted state, but later came to view it "as a matter of the very distinction between inside and outside."[36]

It is important to note that Ngai's analysis primarily concerns a specific manifestation of the anxious subject in Western intellectual history: the male philosopher whose gendered subject position is under threat, catalyzing a state of anxiety that works strategically to reestablish his gendered position. In this way, Ngai's argument casts anxiety as what we might call a technology of gender, exorcising the female or feminine from the male or masculine and allowing a male knowledge-seeking subject to maintain epistemological authority. (It is for this reason, as Peter Stallybrass and Allon White argue, that the "low-Other" is both "socially peripheral" and "symbolically central.")[37] Similarly, Ahmed has shown how anxiety functions as a technology of race, shoring up the boundaries of a white subject in the academic field of critical whiteness studies. As she writes, "The anxiety about borders works to install borders: whiteness becomes an object through the expression of anxiety about becoming an object."[38] In Ahmed's and Ngai's accounts, anxiety is thus theorized as a method to resolve endangered subject positions through projection; it is a technique of boundary management.

These accounts depart significantly from popular psychological approaches, insofar as Ngai and Ahmed theorize anxiety as a solution to a problem, rather than a problem in search of a solution. Incidentally, this helps to explain the Western philosophical romanticization of anxiety as an indication of literariness and acute intelligence, or as an integral step in the acquisition of self-knowledge. For example, as David H. Barlow writes (following Kierkegaard), "Anxiety . . . is rooted not just in a fear of death, but in a fear of nonbeing, nonexistence, or nothingness. Only through recognizing and confronting this fear of becoming nothing—only through the threat of dissolution of the self—can one truly discover the essence of being. Only through this experience can one achieve a clear distinction of the self from other objects or from nonbeing."[39] For Kierkegaard, anxiety is thus a useful, even necessary experience that allows the self to understand itself as a self, indeed to understand the essence of being. This leads Ngai to the insight that the philosophical romanticization of anxiety is rooted in the "male subject's quest for interpretive agency"—an insight that helps to explain Kierkegaard's somewhat puzzling celebration of anxiety as "the dizziness of freedom" and a "desire for what one fears."[40]

Following Ngai's and Ahmed's arguments, it would be a misstep and oversight to take at face value the concerns about work examined in the chapters that follow. Redescribed as expressions of anxiety, these concerns can be better understood as part and parcel of an attempt to reinforce or reestablish valued forms of relationality through producing, performing, or determining the boundaries that delimit a responsible, collectively oriented subject. "Boundary," here, is not meant physically, as in Ahmed's discussion of sensation, but rather refers to the qualities that constitute a subject in contrast to its other, for example the white subject (in Ahmed's argument) or the male subject (in Ngai's argument).

As in Ngai's account of anxiety as a technology of gender and Ahmed's account of anxiety as a technology of race, contemporary anxieties about the transformation and dissolution of work as a social institution project a series of concerns that serve as a ground against which properly bounded subjects can be established. As Barad asserts, boundaries between objects do not exist outside of their conditions of measure such that an incursion can simply be empirically observed. Rather, the description of these boundaries and their dissolution participates in the determination of that which they claim simply to describe. Concerns about the transformation and dissolution of work are thus productive, even in the absence of any corrective action, insofar as they establish conditions for determining the responsible subject under threat. In this vein, the concerns examined in the chapters that follow are not only or even primarily aimed at improving conditions of work, but rather at disciplining the irresponsible, lazy, hedonistic, unruly, promiscuous non-subject and reforming a properly bounded, responsible, collectively oriented subject through the anxiety of the text.

## Policing the Solitary Pervert

In a particularly striking passage of *Phenomenology of Perception*, Maurice Merleau-Ponty describes a disoriented subject, drawing from Max Wertheimer's "Experimentelle Studien über das Sehen von Bewegung" (Experimental Studies on the Seeing of Motion): "If we so contrive it that a subject sees the room in which he is, only through a mirror which reflects it at an angle at 45° to the vertical, the subject at first sees the room 'slantwise.' A man walking about in it seems to lean to one side as he goes. A piece of cardboard falling down the door-frame looks to be falling obliquely. The general effect is 'queer.'"[41] In her essay "Orientations: Toward a Queer Phenomenology," Ahmed seizes upon this image, and on Merleau-Ponty's use of the word "queer." She takes the word both as Merleau-Ponty means it—to describe the wonkiness of being out-of-alignment—but also in its contemporary sense as a catch-all for "nonstraight sexual practices."[42] Putting these senses of queer together, Ahmed extends Merleau-Ponty's analysis to make a point about how "disorienting" it is for queer people to be located in straight spaces, again with a play on words: dis/orientation in the sense of sexual orientation but also of belonging. The disorientation of being queer is thus a function of not being in alignment, not belonging, and not following norms. Ahmed elaborates:

> The normative can be considered an effect of repeating bodily actions over time, which produces what I have called the bodily horizon, a space for action, which puts some objects and not others in reach. The normative dimension can be redescribed in terms of the straight body, a body that appears in line. Things seem straight (on the vertical axis) when they are in line, which means when they are aligned with other lines. Rather than presuming the vertical line is simply given, we would see the vertical line as an effect of this process of alignment. Think of tracing paper. Its lines disappear when they are aligned with the lines of the paper that has been traced: you simply see one set of lines. If all lines are traces of other lines, then this alignment depends on straightening devices, which

keep things in line, in part by holding things in place. Lines disappear through such alignments, so when things come out of line with each other the effect is "wonky." In other words, for things to line up, queer or wonky moments are corrected. We could describe heteronormativity as a straightening device, which rereads the "slant" of queer desire.[43]

In this passage, norms—here heteronorms, though Ahmed also discusses homonormativity in the essay—are theorized as a kind of "straightening device," aiming to bring that which is queer and wonky into alignment. These norms recede or disappear from view when things are straight and in line. Again, "queer" is used here both to describe the out-of-line—drawing from the Greek root of the word, which refers to the "cross, oblique, adverse"—as well as the non-straight.[44] The coincidence of these two meanings is no coincidence, brought together in the "odd, bent, twisted" characteristic of non-normative sexualities which renders them queer.[45] To put it another way, queer sexual orientation is disorienting precisely in relation to heteronorms (and, perhaps, homonorms).

Ahmed's argument is useful for its description of how norms work both productively (putting some objects into reach) and restrictively (keeping other objects out of reach), for its expansive understanding of queerness, and for its account of how norms recede from view for those aligned with them. But it is important to note that Ahmed's argument is but one iteration of a primary thread in queer scholarship, which draws into question norms and normativity as agents of social control. As Robyn Wiegman and Elizabeth A. Wilson propose in their introduction to a 2015 special issue of *differences* ("Queer Theory without Antinormativity"), "Normativity [is] queer theory's axiomatic foe."[46] Wiegman and Wilson cite a number of scholars whose work has been particularly central in advancing antinormative arguments, including Leo Bersani, Judith Butler, Michel Foucault, Gayle Rubin, Eve Kosofsky Sedgwick, Michael Warner, Lee Edelman, Judith Jack Halberstam, Lisa Duggan, Jasbir Puar, and Roderick A. Ferguson, with an additional nod to feminist theory, women of color feminism, and transgender studies. Wiegman and Wilson note that the field's commitment to antinormativity shapes its approach to its objects, from colonialism, war, and incarceration, to disability, affect, and the post-human. They write, "Normative sexualities, normative genders, normative disciplinary protocols, normative ideologies, normative racial regimes, normative political cultures, normative state practices, and normative epistemes: these figures of normativity have been at the heart of queer theoretical inquiry for nearly three decades."[47] While Wiegman and Wilson have been taken to task (most forcefully, in my view, by Halberstam) for falsely representing a common antinormative commitment in queer scholarship that does not exist, thereby simplifying the nuance and diversity of the scholarship they reference, they nevertheless highlight the centrality of thinking about norms and normativity in this body of work.[48]

In fact, Halberstam has elsewhere also highlighted the centrality of (anti)normative thinking. In an introduction to a 2005 special issue of *Social Text*, Halberstam, José Esteban Muñoz, and David Eng note the importance of "subjectless" critique to queer theory—that is, the notion that there is no queer subject; to be queer is to be exterior to processes of identification and subjection—and the way that this critique reveals normalization to be a site of violence.[49] Citing work by Michael Warner, they observe that the normal is produced through identifying the pathological, and that this process cannot be interrupted by simply tolerating or politically representing normalized "queer" subjects; it can only be interrupted through resisting the normal and the normative. This is also to say that the queer is not simply opposed to the hetero but, perhaps more fundamentally, to the normal/normative, which is constituted in part through the social regulation of sexuality.[50]

That said, should we take the publication of "Queer Theory without Antinormativity" as a sign that antinormative critique is on its way out, the inevitable backlash against the backlash? Wiegman and Wilson's critique-of-the-critique targets queer scholars' insistence that norms are restrictive and exclusionary (rather than productive). As they explain, "By transmogrifying norms into rules and imperatives, antinormative stances dislodge a politics of motility and relationality in favor of a politics of insubordination."[51] Collapsing the statistical sense of "normal" (as in average) with its regulatory sense (as in the good or ideal), Wiegman and Wilson suggest that norms cannot be exclusionary precisely because they measure the entire population: in the computing of averages, outliers are just as important as those closest to the middle. "The center," they write, "calls on and is constituted by the periphery"—an uncontroversial proposition for those familiar with post-structuralist thought—rendering the distance between center and periphery "nonsense."[52] To draw an analogy, norms here seem to work like capital: ever-expanding, dynamic, and transformative, rather than prohibitive, denying, and exclusionary. Opposition, in this framework, becomes not only misguided, but impossible.

In a related vein, Wiegman and Wilson make the provocative claim that "lifeless norms" are in fact imagined by scholars so that they might self-form as good (where norms are bad), just (where norms are unjust), and so on. While Wiegman and Wilson's critique focuses primarily on what they see as scholars' misunderstanding of what norms are and how they work—a bold assertion given the size of the field they are targeting—their critique seems motivated not simply by the will to correct this misunderstanding, but by a political sensibility that rejects the oppositional stance critics have taken. This opposition to opposition is understandable (though ironic) insofar as opposition weds the critic to the thing opposed, albeit in an inverted form; bad norms are replaced by better norms. In fact, this is a rare point on which Wiegman and Wilson and some of their critics seem to agree.

This may be true of most norms, except the norms I am concerned with here, which target valued forms of relationality, which is to say valued ways of relating to others. This is because norms, like values and ethics, are socially inscribed; an "opposition" to the social entails no inversion but rather only negation. Insofar as opposition always implies some kind of counter-affirmation—good norms to replace bad norms, a just state to replace an unjust state, a solution to follow critique—a concept like negation (or "queer negativity" as it has been called) is necessary to imagine an "outside" to the social. As Lee Edelman observes, opposition is integral to the proper functioning of the social. On this point he cites Adorno: "Society stays alive, not despite its antagonism, but by means of it."[53] Rather than identifying a new good, Edelman thus calls for "an end to the good as such."[54]

Despite the commitment of queer theorists to thinking critically about normativity (if not adopting some kind of antinormative position), few thinkers are, like Edelman, interested in applying the critique of normativity to those norms that work to reproduce the social itself, though such norms would seem to be the norm's purest expression. For example, Lisa Duggan dismissively notes in her critique of the *differences* issue on antinormativity that "[the antisocial] paradigm went out by 2002 (in the queer studies 'field' that I read), and the withdrawal from the social characterizes only a tiny archive at this point."[55] It seems that for many queer thinkers, it is not norms themselves that are an issue, but rather particular norms. Even those writing about risky and/or anonymous sex have found ways to turn seemingly non- or anti-normative practices into an ethics lesson, as in Tim Dean's defense of barebacking (the practice of having unprotected anal sex) as an ethical practice of forging intimacy with a stranger—a practice whose value to civil society and democracy Dean makes quite explicit.[56] (That Dean leans on Jane Jacobs's *The Death and Life of Great American Cities* in his defense of barebacking should be a giveaway that an ethics lesson is forthcoming.) Dean also cites Samuel Delany's *Times Square Red, Times Square Blue*, another surprising argument for the value of public sex venues as democratic institutions, as if porn theaters and gay bathhouses were as civically oriented as any bowling league or PTA meeting. If there is an effort in these works to avoid pathologizing risky and anonymous sex, this effort is nonetheless mitigated by a valuing of responsibility to others, that bedrock of sociality. Indeed one might draw into question the embrace of pleasure in queer scholarship that otherwise insists that queerness be understood "as collectivity," as Muñoz has written.[57] Here we might ask: Why must pleasure be attached to collectivity? Or better stated: Of what use is pleasure if it must be earned through being a good communitarian subject?

To take another example, in *Cruising Utopia*, Muñoz recalls the social unrest that followed the murder of gay college student Matthew Shepard in 1998. While this unrest must be read, at least in part, in relation to Shepard's whiteness and normative attractiveness (that is, his "telegenic face"), Muñoz recalls that "we nonetheless seized the moment and took to the streets, not only for Shepard but for the countless women and men of all colors who have survived and not survived queer violence on the streets of New York City and elsewhere."[58] As Muñoz recounts, these protests allowed queer people to see themselves as a powerful mass—as having power that "can be realized only by surpassing the solitary pervert model and accessing group identity."[59] Once again the solitary pervert is found lacking, insufficiently socialized, incorrectly oriented toward the collective—a critique not so distant from more explicitly conservative condemnations of homosexuality as inimical to the repro-duction of the social. On this point we might recall a reminder issued by Michael Warner: "The history of the movement should have taught us to ask: whose norm?"[60]

It is in this context that I find particularly useful (and rare) as an interrogation and critique of valued forms of relationality the antisocial thesis in queer theory, presumably the referent of Duggan's dismissive comment. (It is also in this context that I find Wiegman and Wilson's move away from antinormative critique premature.) Often associated with Edelman (despite his rejection of the label), this line of argument is typically thought to originate with Leo Bersani's 1987 essay "Is the Rectum a Grave?" with its fierce critique of attempts to pastoralize "gay sex" and thereby normalize queer populations, as well as its claim that the "inestimable value of sex" lies in its "anticommunal, antiegalitar-ian, antinurturing, antiloving" qualities—an unquestionably provocative argument at a time when other academics and activists were working to combat the stigma surrounding gay sex in the context of the AIDS epidemic by asserting the respectability of the gay "community."[61] (In-deed, the popularity of the label "community" seems less an empirical description than a normative aspiration.)[62] The antisocial thesis would be further developed in Bersani's 1995 book *Homos*, particularly its last chapter, "The Gay Outlaw," which begins with the question, "Should a homosexual be a good citizen?"[63] Connecting Bersani's argument with Ahmed's argument addressed above, I aim ultimately to show how the anxieties examined in the chapters that follow work in the service of norms that prescribe proper forms of relationality—the others to Ber-sani's list: communal, egalitarian, nurturing, and loving. Using Ahmed's language, these anxieties can be understood as a kind of straightening device, engendering the formation of properly social subjects (as op-posed to antisocial non-subjects).

While the object of my analysis here is not sexuality, but rather anxi-eties about the transformation of work, it is through an interpretation of homo desire that Bersani comes to theorize the social as a prison of sorts, and the antisocial as liberatory (an argument that will ultimately allow me to redescribe anxieties about work as serving a pro-social, nor-mative project). It is thus necessary to entertain homo desire as an object of analysis in order to capture fully Bersani's argument. For Bersani, it is not simply that homo-ness resists or disrupts normal socialization, but that *"homo-ness itself necessitates a massive redefining of relationality."*[64] He continues, "More fundamental than a resistance to normalizing methodologies is a potentially revolutionary inaptitude—perhaps inherent in gay desire—for sociality as it is known."[65] (The phrase "as it is known" is important, for reasons I will explore shortly.) The "most radical political potential of queerness," Bersani argues, is to "put into question sociality itself."[66] Proposing "inaptitude" as a political mode, in contrast to "resistance," Bersani suggests that politics need not take the social form prescribed by Muñoz and others.

For Bersani, the social is predicated on a particular psychoanalytic relation to the other, in which the other's difference presents a threat that needs to be annihilated, whether by exclusion, extermination, or assimilation; this is the foundation of his critique.[67] This drive to annihilation is the problem with hetero desire, and the animating force of its normativity. In contrast, homo desire does not simply avoid taking the other as an object; it also and more fundamentally "presupposes a desiring subject for whom the antagonism between the different and the same no longer exists."[68] In other words, homo desire does not simply privilege sameness over difference, insofar as this desire is internally constituted by a difference already incorporated through sexual object choice; gay men, for example, desire in the mold of hetero women (their mothers, at least mythically). It is not that gay men have the "souls" of hetero women—nonsense quickly dismissed by Bersani—but that through homo desire, the different-same binary collapses. As Bersani writes, "Homosexual desire is desire for the same from the perspective of a self already identified as different from itself."[69] In other words, the homo self is different from itself (as male) because it has incorporated hetero woman's otherness as a "major source of desiring material."[70] The homo self is thus both constituted and undone by an originary, incorporated difference.

One might ask here: Why should it matter politically, whether the other is incorporated as a desiring subject, as in homo formation, or a desired object, as in hetero formation? For Bersani, these are fundamentally different: while hetero male desire is "defensive and traumatic," grounded in a "misogynous identification with the father and a permanent equating of femininity with castration," in homo male desire difference has been de-traumatized through an identification with and incorporation of the hetero mother as a desiring subject.[71] So whereas hetero desire is "grounded in lack," and is consequently tasked with and structured by the aim of filling this lack through incorporating the other—what bell hooks has called (in a different context) "eating the other"—homo desire involves a kind of "self-effacing narcissism" that is less interested in the other, and thus less threatened by the other's difference.[72]

In a 2013 interview, Bersani revises this position on "homosexuality as a kind of psychic correspondence of sameness."[73] He explains, "This now strikes me as taking the sameness in same-sex desire too literally."[74] For example, two gay men can of course be more different from each other than a gay man might be from a heterosexual man. For my purposes here, however, the truth value of Bersani's arguments about hetero-ness and homo-ness is less important than are the insights about relationality and sociality he draws from these, especially those insights concerning the radical political potential of a kind of self-centered but also self-shattering escape from the social and fundamental transformation of the relational.

Bersani's reading in *Homos* of Andre Gide's *The Immoralist* provides a sense of what such a transformation might look like. In this reading, Bersani admires the protagonist Michel's "profound indifference" to the otherness of the boys he sexually desires, and for whom he abandons his marriage.[75] Bersani writes that this indifference "means that he demands nothing from them."[76] He continues: "Michel asks nothing more of the objects of his desire than to share a certain space with them; his homosexuality is a matter of positioning rather than intimacy."[77] This is narcissistic in a sense, stemming from a desire to touch an approximate replication or extension of oneself, though without ego; the ego cannot survive this narcissism, which is too implicated in an extensive and "impersonal sameness," as Bersani describes it.[78] The ego dissolves in the act of loving the other as same ("love," Bersani writes, "for want of a more precise word"), which puts at risk the boundary between self and other, providing a foundation for a sort of ethics without the other; an an-ethical ethics.[79] Bersani takes Michel's relation to the boys he desires as a model of how one might "move irresponsibly among other bodies, somewhat indifferent to them, demanding nothing more than that they be as available to contact as we are, and that, no longer owned by others, they also renounce self-ownership and agree to that loss of boundaries which will allow them to be, with us, shifting points of rest in a universal and mobile communication of being."[80] Insofar as the terms "social" and "relational" are sometimes used interchangeably (rather than distinguishing the former as a valued mode of the latter, as I have done throughout this book), the labeling of Bersani's argument as antisocial is somewhat misleading. His project might be better characterized—as he writes—as searching for "an anticommunal mode of connectedness we might all share, or a new way of coming together"; in other words, it is anticommunal but not antirelational.[81] As Tim Dean quips, "Everyone knows that homosexuals throw fabulous parties."[82] Dean explains that the form of desirable relationality imagined by Bersani is one that lies "beyond the normative coordinates of selfhood."[83] It is a promiscuous relationality, a kind of hooking up, an "orgy of connection that no regime can regulate."[84] Whereas the term "social death" has been used to describe the exclusion of the oppressed from the social, perhaps we might think of antisocial relationality as a kind of desirable social suicide.

Bersani has also elaborated this anticommunal mode of connectedness using Georg Simmel's concept of sociability as a kind of relationality void of underlying interests and desires—economic, erotic, or otherwise—but rather motivated by the pleasure of unencumbered contact. As Bersani writes, "Sociability, as the great sociologist discovered, is the one social structure that owes nothing, in its essence, to the sociology of groups."[85] For Bersani, there is pleasure in the rhythmicity of association—"binding and loosening, conquering and being vanquished, giving and taking," in Simmel's words—before the imposition of group identities and their corresponding structuring interests and desires.[86] Sociability is not simply a precursor to the formation of the social, however; it also provides relief from the friction of the social, as well as what Bersani describes as an "intransitive pleasure" produced when we are "'reduced' to an impersonal rhythm."[87] To illustrate what this rhythmic sociability might look like, Bersani considers the practice of cruising for anonymous sex in gay bathhouses. The anonymity of association is important in terms of shedding much of one's individuated personality, but most importantly, Bersani writes, "the intimacy of bodies no longer embellished or impoverished, protected or exposed, by the 'clothing' of both dress and character offers an exceptional experience of the infinite distance that separates us from all otherness."[88]

To reiterate, Bersani's explication of the politics underlying homo and hetero identity is of interest here primarily insofar as it draws into question what appears to be a self-effacing desire to commune with the other—a desire characteristic of the Left—but is actually a desire, motivated by a narcissistic fear of the other, to annihilate the other's otherness through incorporation via the social, squeezing infinite otherness into categorized difference. It is the desire to commune with the other that thus ironically expresses a fear of otherness, and not, as some critics have contended, the indifference toward the other found in various formulations of the antisocial thesis. For example, in a brief but devastating and influential critique, Muñoz suggests that proponents of the antisocial thesis fear "contamination by race, gender, or other particularities that taint the purity of sexuality as a singular trope of difference."[89] It is for this reason that he declares the antisocial thesis to be "the gay white man's last stand."[90] But contamination seems to be precisely the point, as in Bersani's reading of *The Immoralist*. The withdrawal from the social is not a retreat into a kind of individualism that might render one safe from contamination by the other, but rather an undoing of the social/individual axis on which concepts like "contamination" depend for their coherence. The impersonal intimacy that appeals to Bersani entails a profound openness to those with whom we come into contact, an openness that stems from indifference and is, therefore, implicitly "respectful" of difference—or to put it psychoanalytically, lacking an Oedipal agenda—at the same time that it troubles the axis of same/different, in which the different is always in the subordinate position.

Muñoz's critique was effective in putting the kibosh on the antisocial thesis. Once it was associated with a gay, white, male resistance to intersectional thinking, who would rush to its defense, particularly with the contemporaneous rise of queer of color critique? Because of the influence of Muñoz's critique, it is useful to note briefly—as a rejoinder—the resonance of the antisocial thesis with certain strands of postcolonial thought; at the very least, this resonance suggests that the core theoretical assumptions of the antisocial thesis cannot be easily ascribed only to a gay, white, male perspective. For example, the notion that interest in and care for the other—foundational to the social according to Bersani, and central to the ideology of the Left—is motivated by a self-protective annihilative ego "instinct" resonates strongly with Rey Chow's critique of anthropologists' desire to establish the lost or muted subjectivity (or "voice") of the so-called native. Such efforts, Chow theorizes, locate the native's authenticity in a moment preceding the arrival of the colonizer. As Chow writes, "'Subjectivity' becomes a way to change the defiled image [of the native], the stripped image, the image-reduced-to-nakedness, by showing the truth behind/beneath/around it."[91] Importantly, she links this displacement of image for voice (or objectivity for subjectivity) to the critic's desire for control: "Our fascination with the native, the savage, the oppressed, and all such figures is therefore a desire to hold onto an unchanging certainty somewhere outside our own 'fake' experience. It is a desire for being 'non-duped,' which is a not-too-innocent desire to seize control."[92] As in Bersani's theorization of the social, a "fascination" with and, we might add here, assertion of responsibility to the oppressed can thus be understood as masking a desire to seize control behind a veneer of humanitarianism; what appears to be self-abnegation is in fact narcissism, and not the self-shattering "impersonal" narcissism Bersani embraces.

The fascination with the other critiqued by Chow and Bersani is both legitimated and obscured through a notion of responsibility that uncannily inverts the critic's dependence on the other, making it seem as if the critic were prostrate before the other, beholden to the other's needs and deferential to the other's interests, when in fact it is the other who is beckoned to act as the symbolic background against which the critic's subjectivity can be asserted and take shape, entering the unfortunate drama of subject/object relations identified and critiqued by Chow. This notion of responsibility presumes people to be rational agents or subjects with free will, such that they can be said to cause things to happen in the world as an expression (or lack thereof) of their ethics, and thereby be made accountable for their actions. As François Raffoul argues, "Identified with the concept of accountability, responsibility . . . designates the capacity of an agent to be the cause and ground of its acts. The unceasing calls for responsibility in contemporary culture are always calls to such agency, to the position of a subject-cause."[93] To be accountable is therefore not simply to be made to account for one's actions, but to be made to occupy a subject position such that an accounting becomes possible. In this way, the notion of responsibility discursively produces that which it claims merely to describe: the rational, causal agent/subject with free will. It does this, in part, through the policing of what must be labeled, in contrast, as irresponsibility.[94] The policing of irresponsibility is not incidental, but rather is central to the sphere of power and control discursively inscribed through the concept of responsibility.

This insight will prove illuminating in the chapters that follow, which detail a series of continued and strenuous efforts to reconstitute the social and the "subject-cause" that lies at its center, such that our behavior might once again be evaluated according to its proximity to the social good, and managed in relation. Of course, the social good is a variable concept, just as the idealized social responsibility of the Left can be distinguished in several important ways from the idealized individual responsibility of the Right. For the purposes of this book, I will be concerned primarily with arguments that imagine the social good through the Left's notion of collective responsibility, which abnegates a particular class of pleasures understood as self-oriented and self-indulgent, while valuing activities understood as self-abnegating and self-sacrificial. Following Bersani's and Chow's arguments we can redescribe this selflessness as a kind of disavowed self-interest masked by a veneer of altruism, as if efforts to include the other were not motivated by what Chow characterizes as "the surplus value of the oppressed"—that is, the value that accrues to the communitarian, collective subject through "including" the other.[95]

In addition to this affinity with a particular postcolonial perspective, the antisocial thesis resonates with strands of queer scholarship that move beyond a gay, white, male purview. For example, in an article for the special issue of *differences* edited by Wiegman and Wilson, Madhavi Menon makes an argument for a kind of universalism that easily reminds of Bersani's embrace, via Plato's *Phaedrus*, of what he calls "universal singularity" over and above "psychological particularities" and "personal difference."[96] For Menon, as for Bersani, universalism does not partake of the Enlightenment fantasy of an absence of difference, but rather constitutes an "indifference to difference."[97] This kind of universalism does not aim to eliminate particularities; instead, it opposes the formation of identities based on these particularities, so that—for example—one might be Hindu or Muslim without being *a* Hindu or *a* Muslim, as Menon discusses. Insofar as the social is predicated on the formation of coherent identitarian subjects, this universalism of "noncohering particulars" resonates with the antisocial thesis, in which the particulars of identity always aspire to a kind of wholeness that ironically insists on the eradication of its constitutive other.[98]

Contemporary critiques of the antisocial thesis as insufficiently radical or "political" if not surreptitiously reactionary (racially or otherwise), are somewhat surprising given queer theory's commitment to questioning normativity. Sara Ahmed's formulation of the queer as wonky or out-of-line is again useful. As Ahmed describes heteronormativity as a straightening device, bringing the queer body back into line, so too might we think of responsibilization—including through the valuing of community in queer scholarship—as a straightening device, especially insofar as it is the aversion to collectivity, whether in the Right's formulation (family, nation) or the Left's (community), that lies at the heart of queerness, as the antisocial thesis suggests. To make responsible is to reorient or straighten a queer non-subject out-of-line with the social, to insist on an identification with and through the social as a way of rectifying this queerness. Even the Left, with its profound reservoir of tolerance and compassion, cannot seem to shake a subdued but intractable antipathy to forms of queer culture that take pleasure in the superficial, the excessive, the frivolous, and the decadent. If the word "cadence" describes a kind of rhythm—a coming together of disparate parts into synchronicity—"decadence" might be understood as an undoing of rhythm, a falling out of synchronicity with the whole, much as Ahmed's "queer" signifies the wonkiness of being out-of-line with the normal. However, whereas "queer" has gradually become a synonym for nonstraight sexualities, thereby losing some of its critical edge, the charge of decadence, degeneracy, frivolity, excess, and superficiality— the bad others to good, responsible forms of relationality—still stings.

The arguments examined in the chapters that follow suggest that the "solitary pervert" Muñoz finds lacking—taking this figure as a discursive catch-all for the antisocial non-subject (though solitude, as Bersani and Dean suggest, is not the point)—far from being politically inert, is quite a troublemaker. These arguments are unintentionally very instructive, politically speaking. As Ranajit Guha has argued, the presence of insurgent consciousness can be "affirmed by a set of indices within elite discourse," whose "words, phrases and, indeed, whole chunks of prose . . . are designed primarily to indicate the immorality, illegality, undesirability, barbarity, etc. of insurgent practice and to announce by contrast the superiority of the elite on each count."[99] These words, phrases, and chunks of prose, Guha writes, "have much to tell us not only about elite mentality but also about that to which it is opposed."[100] Similarly, Lee Edelman has suggested that we "listen to, and even perhaps be instructed by, the readings of queer sexualities produced by the forces of reaction."[101]

In this spirit, the anxiety surrounding the transformations of work described in the chapters that follow can be understood as part and parcel of an effort to neutralize the threat embodied by the solitary pervert, calling this figure—and, by proxy, the reader—back to responsibility and collective, communal forms of relationality—away from decadence and back to cadence—inadvertently revealing the transgressive character of refusing or, to put it more passively, simply being unavailable for social relations. In this way, the antisocial thesis and critical theorizations of anxiety can be brought together to demonstrate how anxiety functions in the service of a normative project that aims to responsibilize readers by establishing discursive conditions through which responsible subjects can be asserted. In Althusserian terms, we might say simply that the anxiety of these texts hails a responsible subject.

# 2

# Playing

Your business is our pleasure.
—Corporate proverb

In 2000, on the day before Christmas, the *New York Times* published an opinion piece announcing that the dot com bubble had officially burst. "What a difference a year makes," the *Times* wrote. The Dow Jones Internet Index had fallen 72 percent since March (when the bubble had actually burst, by most accounts), and stock prices for companies like Priceline and Pets.com were plummeting.[1] Ten years and one global recession later, fears of another Internet bubble began to circulate in the press, with headlines like "Is This the Start of the Second Dot Com Bubble?" and "Investing Like It's 1999."[2] As in the late 1990s, these fears were stoked by suspicions that stock in Internet companies was overvalued in relation to these companies' profit potential. High profile, high cost acquisitions like Facebook's purchase of Instagram in April of 2013 (for $1 billion) and Whatsapp in February of 2014 (for $19 billion) brought these fears to the surface.

Something important had changed since the 1990s, though. With the development of social media (or web 2.0), Internet companies had identified a source of economic value on which to build their business models: the participation of users. As Tim O'Reilly and John Battelle—founders of the Web 2.0 Summit, a high-profile tech/Internet industry conference—observed, the companies that survived the dot com bubble were those that did not simply offer preexisting services via the Internet, but rather built applications "that literally [got] better the more people use[d] them, harnessing network effects not only to acquire users, but also to learn from them and build on their contributions."[3] Value, in these cases, "was facilitated by the software, but was co-created by and for the community of connected users."[4] As O'Reilly and Battelle conclude, "Web 2.0 is all about harnessing collective intelligence."[5] More precisely, from a business perspective web 2.0 is all about competing to harness and monetize collective intelligence. Because of network effects, this competition often has high, winner-takes-all stakes, as with Google, for example, which became dominant in the field of search in part because the efficiency of its search algorithm improves with increased use.

While the notion of "cyberspace" now seems rather quaint, one might understand competition between web 2.0 companies in quasi-spatial, colonial terms, with the Internet as a geographic territory whose valuable natural resources—its users—have become the subject of competition between companies. This is precisely how O'Reilly and Battelle framed "the battle to dominate the internet economy" in their introduction to the 2010 Web 2.0 Summit, organized under the theme "Points of Control."[6] In order to better understand competition between companies, O'Reilly and Battelle wrote, the conference would aim to "map strategic inflection points across the Internet landscape."[7] "Map" here is meant both figuratively and literally; the organizing visual motif of the conference was an illustrated map with major commercial web and telecom interests represented as countries or empires fighting for territory. This map was inspired by board games like Risk and Stratego and the fanciful maps of novels like J. R. R. Tolkien's *The Hobbit* and *Lord of the Rings*, as well as by what Rudyard Kipling called "the Great Game" in the late Victorian era—that is, the struggle between Russia and England for control over access to India through Afghanistan and "all the stans," with the idea that "if you control the passes you control access for armies, access for commerce."[8] While politically and socially obtuse—not surprising, coming from Silicon Valley—these maps highlight the extent to which much of the value at stake in controlling Internet "ecosystems" is user-created, co-created, or, at the very least, user-dependent.

To be clear, it is not simply market share that companies are competing for; the users of the services and applications in question are not simply consuming a preexisting product but rather are participating in the very production of this product. This participation can take various forms. Some of these forms—blogging, uploading images, tweeting, or producing content in some other way—are intentional: users knowingly and actively contribute some form of data, even if they do not know how this data may be collected or used. Activities like entering strings of text when searching the web can also be understood as intentional. Other forms of participation are less intentional or are even unintentional. As the physical world is increasingly networked—a phenomenon O'Reilly and Battelle referred to in 2009 as "web squared" (though "the internet of things" would ultimately become more popular)—users generate large quantities of data simply by going about their daily lives in highly censored, tracked, and surveilled environments. As O'Reilly and Battelle explain, "The Web is no longer a collection of static pages of HTML that describe something in the world. Increasingly, the Web is the world—everything and everyone in the world casts an 'information shadow,' an aura of data which, when captured and processed intelligently, offers extraordinary opportunity and mind bending implications."[9] This information shadow emerges from practices like web browsing; moving through surveilled public spaces; using smartphones' cameras, global positioning systems, microphones, accelerometers, and gyroscopes; enrolling in school; visiting a doctor; driving; and using a credit card.

Once quantified and digitized, the data produced through these various forms of intentional and unintentional participation can be managed to identify and extract value. This practice has been called "information banking," especially in conjunction with cloud computing (whereby data storage and processing is outsourced to third-party businesses like salesforce.com). Like banks, these businesses add value by analyzing data for inefficiencies that can be eliminated and opportunities for growth that can be exploited. While the term "big data" has been used to refer to the enormous data sets now produced through participation in a networked society, it seems increasingly appropriate to use it in the same way as "big pharma," that is, to describe the industry not the product.

How have these transformations, particularly the emergence of user participation as a source of economic value, been understood by scholars of media and the Internet working in various critical traditions? In 2000, as the dot com bubble was collapsing, *Social Text* published Tiziana Terranova's "Free Labor," an essay that proved influential for scholars looking to make sense of user participation within the context of "communicative capitalism," though at the time of its publication, many of these forms did not yet exist; Twitter, YouTube, Wikipedia, Facebook, MySpace, and even the early social networking service Friendster had not yet been invented. The essay takes aim at what Terranova sardonically refers to as the "idealistic cyberdrool of the digerati"—as expressed, for example, in Kevin Kelly's notion of a "self-organizing Internet-as-free-market."[10] Inspired, in part, by the unrest of the "NetSlaves," including seven former AOL volunteers who asked the Department of Labor to investigate whether they might be owed back wages for their work, Terranova's essay offers a counter to the "glamorization of digital labor," arguing that various kinds of voluntary and pleasurable activity central to the vitality of the Internet—"building Web sites, modifying software packages, reading and participating in mailing lists, and building virtual spaces on MUDs and MOOs"—constitute a form of unpaid, exploited labor.[11] Or more to the point, Terranova argues that these activities blur the boundary between labor and leisure.

Some of the activities in Terranova's list—web and software development, content production—were already legible as forms of labor; in addition to building websites, modifying software packages, or producing cultural content on a voluntary basis, one could find formal employment doing these things. It was thus neither particularly innovative nor controversial to conceptualize amateur content producers, volunteer message board administrators, or hobbyist programmers as unpaid and exploited laborers, especially considering the extent to which employed workers had been displaced by volunteers, for example at Netscape, where layoffs and an increased reliance on volunteer contributions went hand in hand. And while the voluntary and pleasurable nature of these activities would give some scholars pause, most have agreed that the replacement of paid labor by volunteer activities is exploitative in some way.[12] The logic behind this transformation is fairly transparent: corporations profit when content creation is crowdsourced rather than managed in-house, in addition to saving themselves the headache of managing creative-class workers, as Andrew Ross has argued.[13]

The more provocative move in Terranova's essay was to consider as labor other kinds of voluntary and "pleasurably embraced" activity—accessing websites, chatting, and participating in mailing lists—which are not already legible as forms of labor.[14] Unlike voluntary web or software development, these activities displace no jobs, rarely warrant or require remuneration, and are thus not intuitively understood as work-like, yet Terranova suggests that reframing these activities as work, or as a hybrid of work and leisure, helps to draw into focus how economic value is created and captured online, and thereby to illuminate broader political-economic tendencies within contemporary capitalism. Terranova's argument soon proved prescient with the emergence and eventual industry dominance of social media in the 2000s, which occasioned a widespread rethinking of inherited theorizations of labor and leisure, and sparked a debate about whether activities that appear to be leisure should be reconceptualized as labor or labor-like, or, to put it another way, whether experiences of leisure mask various kinds of hidden labor.

This chapter focuses on this debate as it has been expressed in relation to two "objects": first, what Julian Kücklich influentially termed "playbor," that is forms of voluntary, pleasurable online activity that are not typically understood as labor but that are similarly productive of economic value; and second, "creative class" or "no-collar" employment, and in particular the informal design, relaxed norms, and unusual amenities such as video games, skateboards and scooters, volleyball courts, haircuts and massages, free food, and so on, often associated with Internet company campuses and workplaces. As with critiques of playbor, scholars have argued that encouraging certain forms of leisure-at-work facilitates the extraction of economic value from creative labor while also obscuring this exploitation from creative laborers. In the first case, scholars redescribe play as work. In the second case, scholars directly relate leisure-at-work to the increased productivity and exploitation of workers. In both cases, what appears to be play or leisure is revealed to be work or fundamentally work-like, or to serve directly the institution of work.

This chapter offers a novel interpretation of criticism of playbor and leisure-at-work. Through analysis of scholarship in these two areas, including texts by Terranova, Ross, Trebor Scholz, Mark Andrejevic, David Hesmondhalgh, and Christian Fuchs, it argues that diagnoses of exploitation do not aim to recapture or withhold the economic value siphoned from users/workers, but rather constitute an effort to devalue certain forms of leisure, especially those that might be classified as irresponsible or antisocial entertainment or "cheap amusements" (to borrow phrasing from Kathy Peiss).[15] Popular criticism of Internet use provides a roadmap of these forms: consuming pornography, endless web surfing, taking and sharing selfies, gaming, and so on. In other words, certain forms of leisure are compared to, conflated with, or linked to exploited labor because scholars reject their pleasures as equivalent in value to the profits they produce, in such a way that users/workers can be seen as underpaid, ripped off. The fault, however, is not placed strictly onto corporate exploiters, since users/workers are complicit in their own exploitation; the primary problem for scholars is that in the first case, Internet users do not see any need for payment—they gladly exchange their data for pleasure—and in the second case, workers who play volleyball or get a massage on company time are seduced into identifying with their employer in such a way that makes them less likely to demand a just workplace with reasonable hours and job security, which is seen as more valuable than those pleasures.

Furthermore, the chapter finds that diagnoses of exploitation also constitute an effort to value collective forms of ownership, control, and management. Even those scholars who do not directly target forms of leisure understood as irresponsible or antisocial still suggest that users/workers need to be made responsible through collective forms of ownership, control, and management, typically because these forms are thought to be essential to the realization of users'/workers' "best interests." Here, then, it is less the presence of leisure/play/fun that is a problem, than the absence of collective ownership, control, and management, though I will ultimately suggest that this amounts to the same thing.[16]

The argument of this chapter is informed, in part, by an inconsistency in the playbor literature: the redescription of leisure as labor defines this new kind of labor as "*a human activity sometimes undertaken solely for pleasure that has economic and symbolic value and can be performed at any time*," as Scholz has written.[17] Yet not all forms of economically or symbolically valuable pleasurable activity raise flags for critics. To say that such activity comprises an expansive category would be an understatement. It might include, for example, all forms of leisure that require payment or investment, whether for goods (such as books, games, sports equipment) or services (theatre, live music, theme parks). However, critics do not target forms of leisure that are paid for with money; they care only about activities that are paid for with time and attention/activity, though this distinction would seem to matter little. For example, there is a negligible difference between paying to see a movie on demand and paying with time and attention (to advertisers) when the same movie is on broadcast television: Are you a consumer if you pay for the movie, but a worker if you watch it for free? This raises the question: What motivates the analytical separation of these two forms of leisure, a separation central to theorizations of playbor or "digital labor" (as Scholz defines it above)?

To be certain, there can be political utility in redescribing as labor activities that are not conventionally understood as labor. Doing so might draw attention to structural conditions that coerce participation. As David Hesmondhalgh points out, this strategy has been employed by feminist scholars in arguing that domestic childcare and home maintenance ought to be considered and recompensed as labor.[18] Terranova's argument in "Free Labor" is crafted in this mold, at least in part; she observes that an overemphasis in media scholarship on open source programming as unpaid labor, and an underemphasis on the labors of browsing or chatting, is evidence of a masculinist bias.[19] But unlike the labors of childcare or home maintenance, leisure online is rarely structurally coerced. This raises another question: What could be the purpose of describing uncoerced, pleasurable activities as labor or labor-like?

The argument of this chapter is also informed by an inconsistency in the literature that critiques leisure-at-work, particularly Andrew Ross's book *No Collar: The Humane Workplace and Its Hidden Costs*, which exemplifies this line of argument—namely, that permitting and encouraging leisure-at-work mask workers' exploitation. Yet workplaces—including creative-class workplaces—routinely offer benefits that escape scrutiny, including health care, retirement packages, and family friendly policies. Where is the outrage, for example, that some universities offer faculty and staff tuition benefits, both for themselves and their children? This raises yet another question: What motivates the analytical separation and critique of particular kinds of workplace perks or characteristics that involve having fun, goofing off, and taking it easy?

Rather than engaging scholarship on playbor and leisure-at-work empirically—for example, by weighing in on how the organization of labor and leisure has changed over time—I see these inconsistencies as requiring interpretation to reconstruct their underlying motivations and aims. This is also to contest the framing of these various strands of inquiry as simply empirical and theoretical, rather than political or polemical.[20] In interpreting concerns about playbor and leisure-at-work, the chapter argues that the devaluing of particular forms of leisure (and conversely, the valuing of work) is motivated not simply by a cultural elitism that favors certain forms of leisure over others but also, beneath this, by an underlying identification and rejection of certain forms of leisure as self-indulgent and irresponsible, rather than by the unremunerated production of value, or the obfuscation of exploitation, as critics contend. In other words, the reason these forms of leisure are not seen as equivalent in economic value to the pleasures they produce is that they are not seen as engendering or connected to social value. On the contrary, these forms of leisure are understood as having little to do with the collective, communal, responsible forms of relationality valued by critics, whether because devalued forms are understood as escapist or as otherwise detrimental to valued relational bonds.[21] But because of the intellectual legacy of cultural and identity studies, it has become difficult for critics to malign explicitly the tastes and habits of the masses without coming across as elitist, like the misanthropic Theodor Adorno (on fans of popular music, for instance: "They are not childlike. . . . But they are childish; their primitivism is not that of the undeveloped, but that of the forcibly retarded.")[22] Instead, scholars find fault with certain forms of leisure as exploitative. To draw an analogy: rather than telling us that they don't like our (in)significant others—which would foreground critics' own subjective, evaluative criteria—they tell us that these others are using us. It is precisely the "apparently" free nature of services like Facebook and Google, and the "apparently" liberated nature of creative-class office environments that make possible this critique.

The diagnosis of playbor and leisure-at-work as exploitative functions not only to devalue particular forms of leisure, but, relatedly, to value the institution of work, which becomes symbolically important insofar as it is linked to the formation and maintenance of responsible social subjects invested in forms of collective governance, which might take root if only users/workers would curtail their appetite for entertainment. Beneath this identification of apparently new forms of exploitation lies yet another iteration of the call for the passive masses to liberate themselves (or be liberated) from Plato's cave, and to become active and assert control over their productive and creative activity through the institution of work. This is not just about the money—the economic value appropriated from users/workers—but more significantly, about users'/workers' "relinquishing" of control over processes of production to capital, that is, their failure to be sufficiently collectively oriented. In other words, for these critics, exploitation is the symptom of a more pernicious underlying problem—namely, the alienation of users/workers from processes of production that are necessary not simply for their biological survival, but for their survival as social subjects. Or perhaps it is pleasure that is the symptom, evidencing the failed socialization of a communally identified social subject.

In short, concerns about the collapse of work and leisure are motivated by a rejection of certain forms of leisure as antisocial and by a related call to the social via the institution of work. It is not so much the sacred space of leisure that is under threat, as it becomes increasingly contaminated by its proximity to work. Rather, it is the forms of relationality valued by critics—responsible, accountable, obedient, and sacrificial—secured precisely through work as a repository of meaning, a symbolic object. Far from simply mapping the collapse of work and leisure, analyses of this transformation can be better understood as an expression of anxiety stemming from the perception of a threat to the social—an anxiety that works to devalue targeted forms of leisure, with the aim of (re)producing a properly socialized subject—that is, a subject disabused of the prospect of getting something for nothing. In this way, what appears to be an empirical, analytical, and theoretical project is also, and perhaps more fundamentally, a normative project.

## Play Becomes Work

The notion that playbor is exploitative draws from several preexisting streams of scholarship. First, it draws from political-economic scholarship on television beginning in the 1970s, when the growing dominance of television as a cultural form brought the issue of audiences' economic value to the attention of scholars working in various Marxist or quasi-Marxist traditions, as Patricia Clough has detailed.[23] In particular, Dallas Smythe's 1977 essay "Communications: Blindspot of Western Marxism" invited scholars to take up television as a technology of production (in the Marxian sense) rather than simply as a technology of consumption.[24] Contesting the notion that mass media is primarily in the business of producing meaningful content, Smythe argued that the audience is actually the media's primary product, aggregated through the distribution of content and then sold to advertisers. Or more precisely, Smythe argued that it is "audience power"—a modification of

Marx's "labor power"—that is sold to advertisers. Watching television could thus be conceptualized as a form of work, specifically the work of desiring mass-produced commodities and of legitimating the state.[25]

Smythe's argument was subsequently picked up, debated, elaborated, and amended by a number of scholars, most notably Nick Browne and Sut Jhally, the latter of whom has similarly argued that "the media are our employers"; audiences work by watching commercials and are paid in programming.[26] Also of relevance to theorizations of free labor, though sometimes overlooked, is Richard Dienst's 1994 book *Still Life in Real Time: Theory after Television*, in which Dienst takes issue with Browne's and Jhally's treatment of television as an episodic transaction between economic agents, rather than a broader system. While Dienst recognizes that advertising often pays for television programming, and that clients who hire advertisers believe that advertisements help to increase or at least maintain their business, he argues that television does not require advertising to sustain itself. Rather, advertising cashes in on time already socialized by television; advertising realizes television's investment in viewing time. Following Marx's argument that capitalism places workers under social conditions that allow capitalists to harness and exploit their productive power, Dienst argues that television establishes conditions that allow television networks to harness the productivity of our free time, even though it appears as if television itself (rather than audiences' viewing power) is the source of its profitability. Aiding this illusion, television audiences relinquish their free time willingly, seemingly unaware of the productive power of their viewing.[27] In short, audiences pay for television with their time, though they rarely experience the time they relinquish as valuable. As Clough aptly summarizes Dienst's argument: "It is not, therefore, in reading images and then consuming advertised commodities that the viewer produces surplus value. The viewer produces surplus value when he or she watches, that is, when a unit of viewing time and television image, having already been capitalized, is used up."[28]

Also of particular relevance to contemporary theorizations of playbor is Jonathan Beller's 1998 essay "Capital/Cinema" (a precursor to his 2006 book *The Cinematic Mode of Production: Attention Economy and the Society of the Spectacle*). Though Beller focuses on cinema as opposed to television, the two theses he proposes fit comfortably in this line of television criticism and theory: "(1) cinematic movement is an extension of capital circulation: the cinematic image develops out of the commodity-form; and (2) cinema becomes directly involved in the process of social production and reproduction by occupying human time and converting visual attention to labor-power—in short, the labor theory of value is a special case of what I call *the theory of the productive value of human attention*."[29] In Beller's analysis, it is thus attention in its general form that produces value, not simply the attention that accompanies willful labor. In this way, Beller argues, value can be produced outside of contexts of work narrowly conceived, for example in movie theaters, homes, and even "the brain itself."[30]

Contemporary theorizations of playbor also owe an intellectual debt to scholarship on consumerism, most notably Alvin Toffler's work on prosumption—a portmanteau of "production" and "consumption," and an influential concept for scholars of free labor—and similar concepts (for example, Axel Bruns's "produsage").[31] While the concept of prosumption has been used primarily to describe the displacement of labor onto consumers (as in the pouring of one's own drinks at fast food restaurants, or the scanning of one's own groceries at the supermarket), it has also been used to describe the process of stimulating and satisfying consumers' desires by formally engaging them in the production process—for example, by inviting them to customize products or co-create experiences.[32] As George Ritzer and Nathan Jurgenson have argued, prosumer retail environments do not simply offer consumers enhanced products or services, but surreptitiously extract value from consumers through this engagement.[33] While Ritzer and Jurgenson theorize prosumption as a new phenomenon, it might also be conceptualized as an extension of existing corporate practices of "coopting" non-market cultural activity and selling it back to consumers, as Terranova has suggested.[34]

Finally, contemporary theorizations of playbor have been influenced by the autonomist school of Marxist thought, and in particular autonomist theorizations of immaterial labor—a term that describes "a series of activities that are not normally recognized as work," but that increasingly take on characteristics of work, particularly in terms of value production and extraction, as Terranova writes, citing Maurizio Lazzarato.[35] ("Immaterial" here refers both to the primacy of affective and cognitive skills involved in this labor and to the products produced, which are often informational and cultural commodities.) Antonio Negri, perhaps the most well-known and often-cited scholar in this tradition, contextualizes the shift to immaterial labor as an expression of "real subsumption," whereby "life in all its walks is constituted as productive labor," as Adam Arvidsson notes.[36] This tendency has culminated in what Jonathan Crary calls "24/7 capitalism": a political-economic system in which there is little time for anything other than production and consumption (these are collapsed in Crary's analysis, as they are for many of the scholars cited above).[37]

The notion that user consumption and participation online can be understood as a form of free labor or playbor has thus been influenced by a number of distinct, though related, lines of thought. Like television audiences, prosumers, and immaterial laborers, Internet users are refigured by critics of playbor as a kind of worker, producing value through activity that the user experiences as leisure, but that capital treats as labor or labor-like. The idea that the boundary between work and leisure is blurring, shifting, or dissolving in this way is thus not entirely novel. Nevertheless, the application of this idea to make sense of user participation has sparked renewed interest and considerable debate.

While scholars largely agree that leisurely consumptive and participatory practices produce economic value, there is much debate, or rather a kind of hand-wringing self-scrutiny, concerning the pleasurable, fun nature of these practices, which might suggest that those who engage in them are not in fact exploited. As Ritzer and Jurgensen write regarding prosumption, "The idea that the prosumer is exploited is contradicted by, among other things, the fact that prosumers seem to enjoy, even love, what they are doing and are willing to devote long hours to it for no pay."[38] This kind of ambivalence is expressed not only by those skeptical of playbor as a concept, but by its proponents. For example, Scholz asks, "Does it really make sense to think of these activities or the updating and 'liking' of status updates as labor?"[39] Terranova, too, is somewhat troubled by the fact that free labor is "simultaneously voluntarily given and unwaged, enjoyed and exploited," and that the Internet "is always and simultaneously a gift economy and an advanced capitalist society"—the word "simultaneously" suggesting that these characteristics ought not coexist, that they do is surprising.[40]

The fun, pleasurable aspect of playbor troubles scholars because it seems to suggest that playborers are not exploited; exploitation is supposed to be immiserating, not fun. Furthermore, the characterization of playbor as exploitative might easily be read as something of an insult to the exploitation of real (material) labor, which is in fact immiserating. As Hesmondhalgh pointedly asks, "Are we really meant to see people who sit at their computers modifying code or typing out responses to TV shows as 'exploited' in the same way as those who endure appalling conditions and pay in Indonesian sweatshops? Clearly not," he asserts in a rejoinder to those who have characterized the web as a kind of sweatshop.[41] Referencing Smythe's work on the audience commodity, Hesmondhalgh suggests that accounts of playbor may even be unethical insofar as they compare incommensurate forms of exploitation. He sardonically asks whether laborers ought to be paid for sleeping, as this also contributes to their reproduction as laborers. For Hesmondhalgh, demanding payment for forms of labor that seem fairly low in the hierarchy of exploitation should only be done for pragmatic reasons, for example to "redistribute income and/or . . . highlight the ethical problems concerned" in cases where seemingly mild forms of exploitation "[contribute] significantly to broader patterns of injustice and inequality."[42]

Hesmondhalgh takes a particularly strong stance against the idea that playbor is exploitative, but even proponents of the concept of playbor make a point to acknowledge and honor the exploitation of material labor as distinct from and more egregious than the exploitation of playbor. As Andrejevic writes, "It is crucial to recognize the difference between types and levels of exploitation and to prioritize critical response accordingly—just as one might distinguish between different types of material deprivation."[43] Later in the essay he doubles down on this assertion, writing that "it is harder to get worked up about the allegedly exploitative conditions of user-generated content sites than about the depredations of sweatshop labor and workforce exploitation."[44] Scholz, too, suggests that global immiseration ought to take precedence, citing exploited Foxconn workers and enslaved miners in the Democratic Republic of Congo who extract the minerals used to make laptops and mobile phones.[45] Similarly conceptualizing playbor as the final moment of a longer value chain that entails the more severe exploitation of material labor, Fuchs writes that "digital labor is based on the surveillance, blood, and sweat of superexploited labor in economic developing countries."[46]

In these various accounts, the exploitation of material labor is understood as most egregious, insofar as it is most closely tied to immiseration. The exploitation of immaterial labor falls next in line, followed by forms of voluntary participation that have displaced immaterial labor. Last in line (in terms of "real" exploitation) is the exploitation of voluntary and pleasurable forms of participation not typically understood as labor—that is, playbor. However, and crucially, most of these scholars subscribe to this hierarchy not to shut down the debate surrounding playbor—as Hesmondhalgh does above—but rather to pay their respects to those workers that are most exploited in order to move forward with an analysis and critique of playbor as exploitative.

Many of the critics who characterize playbor as exploitative use the term "exploitation" in the colloquial sense, that is "to express our repulsion when someone makes use of someone else for their own purposes," as Hesmondhalgh notes.[47] In this use, "exploitation" is synonymous with value appropriation and is taken to be a self-evident wrong; it is enough simply to point out the various ways that corporations profit from various kinds of voluntary participation, or as Andrejevic puts it, "a small owner class benefits from the unpaid labor of the masses."[48] For example, taking a note from autonomists, Scholz writes that "social life on the Internet has become the 'standing reserve,' the site for the creation of value through ever more inscrutable channels of commercial surveillance."[49] Similarly, Terranova notes that "in 1996 at the peak of the volunteer moment, over thirty thousand 'community leaders' were helping AOL to generate at least $7 million a month."[50] The money or value referenced by these scholars is meant to be a smoking gun, evidence of the crime or sin of value appropriation—not only because it is wrong to take that which you did not work for, but also because value appropriation is linked to immiseration, even if not in a direct way.

The notion that voluntary online activity is labor or labor-like and, as such, is exploited (in the colloquial sense of the term), is predicated on rejecting the terms of an exchange. One side has not given enough to render the exchange fair. As Ross writes, "When all is said and done, the informal contract that underpins this kind of economy is a profoundly asymmetrical deal."[51] To assert that some class has been exploited requires a comparison of unequal qualities, in this case, pleasure and economic value. YouTube viewers are entertained; YouTube makes a profit. Critics reject this pleasure as equal in worth to that profit. The pleasure is not sufficient in some way; it is inadequate. Even critics who recognize that there exist forms of compensation other than the wage have described this exchange as unfair. For example, Hector Postigo writes, "While it is true that hobbyists may receive more than just money for their work, when compared with the billions of dollars that video-game companies reap, it would seem that they should gain more than a good reputation for their 'keen' code."[52] The user thus becomes the used in these accounts.

While such characterizations would seem simply to be identifying exploited labor as such, they are actually producing (in a discursive sense) that which they claim to describe. This discursive production requires a prior rejection of the idea that the exchange between companies and users is fair; without this rejection, the exchange would be seen as fair, and the activity in question could no longer be described as labor. As John Roemer has suggested, "Identifying some labor as unpaid often requires a prior diagnosis of exploitation."[53] What I am suggesting here is that identifying leisure activities as labor or labor-like requires a similar prior diagnosis of exploitation (in the colloquial sense of the term) and, behind this, a rejection of the pleasures produced through these activities as equal in value to the profits made by the companies in question.

The phenomenon of "gamification"—that is the application of game characteristics (such as points and badges) to non-game activities—has garnered a line of criticism that exemplifies this rejection. Simply put, critics of gamification contend that users/players have been tricked, accepting insufficient symbolic rewards for the real economic value they produce. For example, McKenzie Wark describes gamification as "getting people to do things without paying them by offering them symbolic rewards in exchange" and characterizes it as a simulation of a gift economy.[54] In a gift economy, Wark writes, "You do something for nothing because you want to do it," rather than for wages.[55] He argues that gamification cannot constitute a true gift economy because "the gift is not to another, and not via another to the commons in general, and the reward is not recognition by others making the same gifts."[56] It is revealing that it is not simply the absence of payment that makes gamification exploitative here, since gift-giving—valued by Wark—also occurs in the absence of any direct payment. Rather, the problem with gamification is that gamified activities are not collectively oriented; it is not only acceptable but admirable to work without payment if this work serves the greater collective good.

Ian Bogost has similarly described gamification as "marketing bullshit, invented by consultants as a means to capture the wild, coveted beast that is videogames and to domesticate it for use in the grey, hopeless wasteland of big business, where bullshit already reigns anyway."[57] The problem, for Bogost (as for Wark), is not simply that "gamified" games are bad games, but that they replace "real incentives" with "fictional incentives" and are thus exploitative; he proposes the term "exploitationware" to describe these games.[58] Bogost extends his critique beyond games to include companies like Facebook and Google, writing that "they use the carrot of free services (their purported product) to extract information that forms the real basis for their revenues (their real product)."[59]

To be certain, Bogost is not wholly opposed to the use of games for marketing purposes. For example, he supports using games to "[help] people understand how specific products and services might benefit particular wants or needs," but he takes issue with "branding and messaging as a way of creating desires through affinity."[60] Bogost's qualm here, it seems, is related to the integrity of social relationships, which he sees as endangered by marketers' abuse of users. As he writes, "When loyalty is real, it's reciprocal. It moves in two directions. Something real is at stake for both parties."[61] Gamification avoids establishing such real relationships, Bogost argues; instead, it replaces these with "dysfunctional perversions of relationships. Organizations ask for relationships, but they reciprocate that loyalty with shams, counterfeit incentives that neither provide value nor require investment."[62] Again, the problem with gamification is not simply that it fails to reward players' work with monetary payment, but that the kinds of relationships it establishes are not "real," which is to say loyal, reciprocal, responsible. The reason these relationships are not real, for Bogost, is that what is given to users in exchange for their participation—fictional incentives—is seen as having false value or no value.

While many, perhaps most, critics of playbor direct their critique at what they see as an uneven exchange, with an underlying notion of exploitation as unjust value appropriation, for some critics—most notably Hesmondhalgh, Andrejevic, and Fuchs—this analysis does not go far enough. For them, exploitation does not simply entail value appropriation but, following a more explicitly Marxian understanding of exploitation, exclusion of the exploited from ownership over productive resources, as well as the dependence of the exploiter on the deprivations of the exploited. As Marxist scholar Erik Olin Wright elaborates, appropriation does not constitute exploitation unless these other two conditions are also met.[63] Similarly, if only these two conditions are met (without value appropriation), the dynamic in question is better described as oppression, not exploitation. While the adoption by critics of this amended understanding of exploitation might suggest that they do not reject superficial pleasures, insofar as they contend that value appropriation does not on its own constitute exploitation, I will argue that the focus on exclusion as an additional condition of exploitation is similarly motivated by this rejection. However, rather than targeting leisure/pleasure directly, they target the absence of collective ownership and control, which similarly amounts to an assertion that users need to be made responsible, whether because collective ownership and control are responsible endeavors, or because the kinds of cultural content produced through collective ownership and control are assumed to be aligned with users' "best"—that is, collective—interests.

Following a Marxian understanding of exploitation, Hesmondhalgh argues that accounts of playbor that equate exploitation with appropriation are flawed ("unconvincing and rather incoherent"); there also needs to be an element of force, however indirect or overarching, given that force is the mechanism of exclusion.[64] Of course, playbor, as defined by all of the scholars cited above, is voluntary, not coerced. Acknowledging this, Andrejevic observes that "coercion does not require someone standing over the worker with a gun or some other threat of force."[65] Rather, coercion can be "embedded in the relations that structure so-called free choices."[66] In other words, users' exchange of personal information for convenience is not free, but rather coerced because this exchange occurs "under conditions structured by the private ownership of network resources," not to mention users' lack of awareness about tracking practices.[67] This theorization of "embedded" coercion allows Andrejevic to argue that users have been forcibly alienated from productive resources, which, in turn, makes possible their exploitation.[68] As he writes:

> The ownership class that includes the founders of Facebook, Google, Yahoo, and so on could not exist without capturing and controlling components of the productive infrastructure. The value that they appropriate stems in large part from their ability to capture aspects of the activity of those who access their resources, and their ability to do so is directly related to their ownership and control of these resources. Bluntly put, the ability to exploit this activity for commercial purposes for the economic benefit of the few would disappear if these resources were commonly owned and controlled.[69]

Following this line of thought, if such resources were communally held and administered, the choice to participate could be reclassified as free.

Users have relinquished control, or have had it taken from them, and this is a more pressing problem than the appropriation of value for critics like Andrejevic, though the former makes possible the latter. As Andrejevic writes, "Crucial resources for interaction are no longer in our hands," adding later that this "deprives workers of control that should be theirs."[70] How to fix this problem? Fuchs argues that the ownership structures of Internet platforms need to be made egalitarian and participatory; only then might Internet use constitute or contribute in any meaningful way to democratic relations. Fuchs imagines a future in which this could happen: "On the communist Internet, humans cocreate and share knowledge; they are equal participants in the decision-making processes that concern the platforms and technologies they use; and the free access to and sharing of knowledge, the remixing of knowledge, and the cocreation of new knowledge creates [sic] and reproduces [sic] well-rounded individuality."[71]

As Fuchs's statement suggests, the argument that users ought to control crucial resources for interaction is about making users responsible through collective ownership. But it also entails an indictment of the "knowledge," or rather the popular culture, produced and circulated online. Fuchs is bolder than most in making explicit his antipathy toward popular culture: "Facebook users are not involved in decisions. Facebook fan groups are dominated by popular culture, with politics being a sideline. Oppositional political figures are marginalized. . . . Facebook is dominated by entertainment. Politics on Facebook is dominated by established actors. Alternative political views are marginalized, and especially critical politics is not often found on Facebook. It is a more general feature of the capitalist culture industry that focuses more on entertainment because it promises larger audiences and profits."[72] Fuchs also refers the reader to a table that lists the most popular Facebook groups (ranging from 50.7 to 34.8 million likes): (1) Facebook, (2) Texas Hold 'Em Poker, (3) Eminem, (4) YouTube, (5) Rihanna, (6) Lady Gaga, (7) Michael Jackson, (8) Shakira, (9) *Family Guy*, (10) Justin Bieber. At the bottom of the list Fuchs includes a few "political" figures for comparison: Michael Moore (495,866 likes), Noam Chomsky (325,325 likes), and Karl Marx (186,722 likes).

In this way, criticism of playbor that follows a modified understanding of exploitation—like criticism that follows a colloquial understanding—remains grounded in the notion that there is a problem with particular forms of leisure, those that Fuchs categorizes as entertainment in contrast to "critical politics." In this case, the problem is more clearly specified: popular culture, it seems, is not aligned with users' best interests, whatever these might be ("well-rounded individuality"?). The attendant argument is that wresting control of the digital means of production would transform cultural production to align with these interests, rather than manipulating users into internalizing desires that are not their own and that work against their interests. That is to say, corporate control is a problem not only because it makes possible the appropriation of value, nor only because it deprives users of control that ought to be theirs, but also because it allows corporations to manipulate users—crafting content to shape their values, knowledge, opinions, tastes, and, ultimately, their behavior—with the underlying assumption that corporations do not have users' best interests at heart, in a way that harkens back to Marx's characterization of alienation as a concentration of control in a "sub-section of the species . . . who then act as gods . . . to direct the trajectory of the rest," as Nick Dyer-Witheford has written.[73] Along these lines, Andrejevic describes the "alienated world envisioned by interactive marketers" as one in which "every message we write, every video we post, every item we buy or view, our time-space paths and patterns of social interaction all become data points in algorithms for sorting, predicting, and managing our behavior."[74] It is not just monitoring or tracking that presents a problem for critics, but managing, or "the systematic use of personal information to predict and influence."[75] Capital is figured as both invasive—permeating the sacred space of the social— and manipulative: "They transform our own activity . . . back upon ourselves in unrecognizable form, servicing interests and imperatives that are not our own."[76]

This indirect problematization of culture is noteworthy for the fact that some critics in this line of thought try to avoid devaluing pleasure. For example, Andrejevic contests the notion that pleasure is at odds with exploitation, arguing that exploitation does not preclude "a sense of enjoyment or pleasure."[77] He elaborates:

> Nor is it the case that accounts of exploitation necessarily denigrate the activities or the meanings they may have for those who participate in them rather than the social relations that underwrite expropriation and alienation. The point of a critique of exploitation is neither to disparage the pleasures of workers nor the value of the tasks being undertaken. To argue otherwise is to stumble into a kind of category confusion: an attempt to reframe structural conditions as questions of individual pleasure and desire. The critique of exploitation does not devalue individual pleasure any more than such pleasures nullify exploitative social relations.[78]

Tellingly, however, the scholars Andrejevic references in support of this claim are Nancy Baym and Robert Burnett, whose work addresses the pleasures of "pre-mass society," "when music was always performed in communities by locals for locals rather than by distant celebrities for adoring fans."[79] Furthermore, in his response to Baym and Burnett, Andrejevic cites workers' pleasure in their craft "or in the success of a collaborative effort well done."[80] So while Andrejevic claims that the critique of exploitation need not devalue pleasure, the pleasures he refers to are all collectively oriented, as if to point out that not all pleasure is bad (selfish, ego-driven, and so on). It seems likely that this is why

Andrejevic wants to avoid devaluing pleasure—that is, to allow for the valuing of collectively oriented pleasures.

The fantasy of a collectively owned and operated Internet imagines the displacement of a commercial relationship between users and service providers—a "bad" relationship where irresponsible pleasures are traded for users' data and its economic value—by a collective, communal relationship along the lines that Fuchs describes above, and which might provide the kinds of pleasure Andrejevic values. If this fantasy has sway, it is because it appeals to a valuing of particular relational forms: the (social) ties that bind. It asks us to invest in the work of collective ownership and operation. It also invites us to imagine the existence of desires that cannot be met through the market—desires whose realization (through the wresting of control over productive capacities) would better serve our interests. As Andrejevic writes, the exploitation of playbor diminishes "the potential of individual and social life"—a potential we are implicitly invited to invest in.[81] What are these desires and interests of "ours," this potential? Whatever they are, this reprieve from alienation—a return to wholeness, the Garden of Eden, and so on—depends on the displacement of fleeting, disloyal, commercial relations by bonded, responsible, social ties.[82] It is not only that the future in which this potential is actualized depends on the social, but that the social must also be the content of this future.

## Work Becomes Play

The other primary site at which certain leisure activities have been critiqued as a form of exploitation is the "creative-class" or "no-collar" workplace, particularly in the new media sector. Though seemingly loosely related—Facebook employees, for example, would be considered in most accounts to be immaterial laborers, while Facebook users would be something like playborers—critiques of playbor and leisure-at-work are related, I will argue, through a similar problematization of particular forms of leisure as self-indulgent and insufficiently collectively oriented. I take Andrew Ross's 2003 book *No-Collar* as exemplary of this line of critique. As I explain below, in Ross's argument this problematization of particular forms of leisure is expressed through a critique of no-collar workers as having prioritized a "humane" workplace over a "just" work-place; in this formulation, the descriptor "humane" is discursively linked to self-oriented pleasures, while "just" is linked to social bonds.

Tech offices and corporate campuses have been the subject of some fascination in the popular press since the dot com boom, with coverage in publications like the *New York Times*, the *Los Angeles Times*, *Time Magazine*, and the *New Yorker*, as well as treatment in popular culture, as in Dave Eggers' 2013 novel *The Circle* and the 2013 film *The Internship*, filmed on the "Googleplex" campus in Mountain View, California. In these various treatments, particular attention is paid to unusual workplace design and amenities, what Ross refers to as "gimmicks." If the playful and autonomous ethos of tech start-ups in the 1990s was exemplified in news stories by the presence of a foosball table, the list of perks now offered at some offices is quite extensive and difficult to capture in a single object. An article in the *New York Times* ("Looking for a Lesson in Google's Perks") lists the following: "a labyrinth of play areas; cafes, coffee bars and open kitchens; sunny outdoor terraces with chaises; gourmet cafeterias that serve free breakfast, lunch and dinner; Broadway-theme conference rooms with velvet drapes; and conversation areas designed to look like vintage subway cars."[83] Allison Mooney, a Google employee interviewed in the article, adds to this list:

> The perks . . . are "amazing." In the course of our brief conversation, she mentioned subsidized massages (with massage rooms on nearly every floor); free once-a-week eyebrow shaping; free yoga and Pilates classes; a course she took called "Unwind: the art and science of stress management"; a course in advanced negotiation taught by a Wharton professor; a health consultation and follow-up with a personal health counselor; an author series and an appearance by the novelist Toni Morrison; and a live interview of Justin Bieber by Jimmy Fallon in the Google office. This in addition to a full array of more traditional employee benefits. Curiously, there's some exercise equipment but no fitness center (Google's headquarters in Mountain View, Calif. has multiple state-of-the-art fitness centers) because Manhattan employees said they preferred joining health clubs to exercising with colleagues. (Google subsidizes the gym memberships.) And there's no open bar, although alcohol is served at T.G.I.F. parties (now held on Thursdays), one of which featured a dating game.[84]

In *No-Collar*, Ross narrates a history of the so-called humane workplace, explaining its evolution as a result (in part) of worker demands—as expressed, for example, in the 1972 strike at the General Motors factory in Lordstown, Ohio. The elements of leisure incorporated into the no-collar workplace might similarly be understood not only as enhancing workers' quality of life, but as an expression of their leverage, that is the power to have their needs and desires met by their employers. In fact, Ross suggests this at several points in the book, arguing that tight labor markets—that is, labor markets in which there are many available jobs and few qualified workers to fill them—generally force companies to compete for workers by offering them higher salaries and better benefits and working conditions.[85] Ross notes that this is precisely what happened after World War II, then again in the 1990s, when some Fortune 500 companies offered perks like concierge services, dry-cleaning pickups, and paid sabbaticals, and most recently in the "New Economy."[86] In short, employers would surely prefer a flooded labor market—that is, a labor market with few jobs and many available workers, which forces workers to compete against each other for jobs and thereby drives down wages, benefits, perks, conditions, and general expectations.

It is significant that despite these observations, Ross seems more interested in critiquing workers' attraction to various perks (when these are prioritized over a "just" workplace) than he is in understanding these perks as an expression of worker leverage, or, for that matter, as simply desirable in terms of quality of life.[87] He makes this turn by noting that establishing a "humane" workplace is often in management's interest as well as workers' interest. As Ross argues, since the 1920s, workplace management reforms have occasioned increased control of employees, in part by courting their loyalty through a humanized work environment, which also diminishes the appeal of trade unionism. This tactic of courting loyalty is especially useful for employers who manage creative labor, which operates rather mysteriously and is thus difficult to control.[88] Because creative labor is unpredictable, the work environment needs to be structured in a way that minimizes employees' need to leave, facilitating productivity as it emerges in fits and starts over long stretches of time and, thereby, the extraction of economic value from creative labor. The notion that workplace amenities are offered not simply in the interest of employees, but to facilitate their productivity is echoed in the *New York Times* article cited above, which quotes Ben Waber, a workplace analyst:

> "Google has really been out front in this field," he said. "They've looked at the data to see how people are collaborating. Physical space is the biggest lever to encourage collaboration. And the data are clear that the biggest driver of performance in complex industries like software is serendipitous interaction. For this to happen, you also need to shape a community. That means if you're stressed, there's someone to help, to take up the slack. If you're surrounded by friends, you're happier, you're more loyal, you're more productive. Google looks at this holistically. It's the antithesis of the old factory model, where people were just cogs in a machine."[89]

For critics like Ross, the problem with workplace amenities like those offered at Google is not simply that they are geared toward the extraction of value from creative labor, but that workers are not even aware of their own exploitation because they have been seduced into identifying with their employers, in part through the "humanization" of the work environment. For example, Ross writes that the Silicon Alley lofts (in New York) inhabited by tech companies

> were reclaimed for industry, but the work they hosted looked more and more like play, and employees were encouraged to behave like artists and keep artists' hours. The Alley's neo-bohemian culture helped sustain the belief that this kind of work was a viable alternative to corporate America. This belief (it may be more accurate to call it a willing suspension of disbelief) was especially important to contrarians with an arts background, who had been trained to scorn the conditions of a middle-class work environment, as well as the routine rhythms of industrial time.[90]

According to this argument, workers have been made complicit in their own exploitation. Allison Mooney (the Google employee quoted in the *New York Times* story above) exemplifies this complicity: "'People want to come in,' Ms. Mooney said. On average, she estimates she spends nine hours a day there, five days a week. She mentioned that she recently took a day off—and ended up at the office. 'I live in a studio apartment,' she explained. 'And I don't have free food.'"[91]

According to critics, this sense of passion and dedication is precisely what management aims to cultivate, and why it has structured the work environment in the way that it has, whether cynically (as a way to exploit workers), idealistically (because management values employees' quality of life), or both. As Ross quotes Craig Kanarick, a founder of Razorfish (an "interactive agency" and the primary ethnographic subject of Ross's book): "If I'm going to go somewhere everyday, it damn well better be a good experience, and if everyone else shows up there everyday, I can't just bribe them. I don't have a crew of really expensive prostitutes here that just get paid to show up. They had to really love what they're doing, and love showing up, and love the culture."[92] Responding to this sentiment, Ross opines, "Love seemed like a high standard to expect of employees and probably not a very healthy one. Even so, I had interviewed dozens of [employees] whose passion for the company probably went beyond the zone of comfort."[93]

This passion seems to disturb Ross; he suggests that these workers have essentially fallen for a trick. As he writes, "It was the social and cultural design of the workplace that stole the affections of employees"—the word "stole" here intimating the unjust or unequal nature of this exchange.[94] Workers have been fooled into believing that management wants what is best for them, rather than to take was is best from them. As the husband of one of Ross's ethnographic subjects tells him, "That's what your book should be about. How the counterculture was duped into thinking they are not working for corporate America. It's like a wolf in sheep's clothing."[95]

However, Ross argues that it is not just the presence of foosball tables that has tricked workers: he emphasizes the importance for Razorfish employees of "grassroots impromptu sport" at the office, over and above "built-in perks" like massages and video games.[96] Ross quotes an employee: "There was the official party line on culture, which was enforced fun, and then there was what we created for ourselves."[97] Of all the aspects of work at Razorfish valued by employees, none ranks higher than the democratic, quasi-anarchist organization of the workplace: "the absence of chains of command between employees and managers," "control over their work," and the distribution of authority and responsibility.[98] Ross writes that employees "experienced the result as a restoration of personal respect and dignity and it was prized as highly as any measure of monetary compensation."[99]

In Ross's argument, this reaction serves to demonstrate workers' desire for autonomy and a less alienated work life, a desire that Ross clearly values. But it is also intimately connected to workers' exploitation. As he writes:

> Features that appeared to be healthy advances in corporate democracy could turn into trapdoors that opened on to a bottomless seventy-hour-plus workweek. Employee self-management could result in the abdication of accountability on the part of real managers and an unfair shouldering of risk and responsibilities on the part of individuals. Flattened organizations could mean that the opportunities for promotion dried up, along with layers of protection to shield employees from market exposure. A strong company culture was an emotional salve in good times but could turn into a trauma zone in times of crisis and layoffs. Partial ownership, or stake holding, in the form of stock options could give employees an illusory sense of power sharing, rudely shattered when they encountered the unilateralism of executive decision-making in layoffs and office closures.[100]

Again, it is not simply that Razorfish exploited its workers in all of these ways—long hours, unfair allocation of responsibility and risk, job stagnation, and so on—but that because of the apparently democratic organization of the workplace, many Razorfish employees, especially those who had worked for the company for a long time, were often unable to distinguish their own interests from those of management, even as they distinguished between top-down and bottom-up culture. Like critics of playbor, Ross thus imagines a kind of bait-and-switch at work here: both the transformation of work (to be democratically self-directed) and the incorporation of leisure into work serve the interests of management, even if these modifications sometimes make it difficult to manage employees in a direct way. Workers, in turn, are seen as suffering from a kind of false consciousness, unable to see their exploitation for what it is.

The most acute manifestation of this false consciousness, Ross suggests, is overwork. As he writes, "When work becomes sufficiently humane, we are likely to do far too much of it, and it usurps an unacceptable portion of our lives. If there is a single argument in this book against the pursuit of the humane workplace, then it rests its case there."[101] However, this issue is made more complicated by the complicity of workers, or more precisely by the transformation of work (or aspects of work) into something pleasurable, perhaps into something that in a certain sense should not even be called work. As Ross continues, "Not by any boss's coercive bidding, but through the seductive channel of 'work you just couldn't help but doing,' had the twelve-hour day made its furtive return."[102] Elsewhere he characterizes New Economy workers as "so complicit with the culture of overwork and burnout that they have developed their own insider brand of sick humour about being 'net slaves', i.e., it's actually cool to be exploited so badly."[103] This raises the question: Can one be overworked if work is pleasurable and not coerced? Or to put it another way: What is the problem with work if not coercion? Why even call it work?

And so it appears that it is not really the (over)work of no-collar laborers that presents a problem for Ross; if work is pleasurable and uncoerced, then overwork is hardly a problem. Nor is the problem precisely that satiated workers make for poor revolutionaries, nor even that New Economy labor is precarious and that today's satisfied workers might be tomorrow's disaffected unemployed or underemployed. Rather, there are two related problems. First, no-collar workers falsely believe that they have more power than they actually have. This lowers their guard, allowing capital to tap into their most creative, personal energy. As Ross writes:

> Perhaps the most insidious occupational hazard of no-collar work is that it can enlist employees' freest thoughts and impulses in the service of salaried time. . . . When elements of play in the office or at home/offsite are factored into creative output, then the work tempo is being recalibrated to incorporate activities, feelings, and ideas that are normally pursued during employees' free time. For employees who consolidate office and home, who work and play in the same clothes, and whose social life draws heavily on their immediate colleagues, there are no longer any boundaries between work and leisure.[104]

The problem with this tapping is not simply that these "freest thoughts and impulses" ought to belong to the individual alone, away from the reaches of capital. One gets the sense that Ross has something else in mind—namely, that workers seem to have both a desire and capacity for collective management, a capacity that is squandered through the sham democracy of companies like Razorfish: "The New Economy was a long way off from a dictatorship of the proletariat. Employee stock option plans offered a stake in company wealth but were rarely linked to genuine forms of employee participation in decisions about policy and work design and almost never to decisions about investments, hiring and firing, and the closure of offices. In this respect, they were a pale shadow of a century's worth of previous attempts, mostly in Europe, at employee self-management."[105] In short, what ought to belong to a collective of workers has been put up for sale in the market (a concern addressed in more detail in chapter 4). For many on the radical Left, the incursion of capital into the sacred space of the community—or, more broadly, into the social—is a serious problem (to put it mildly), in part because it is thought to compromise the sovereignty of labor. When Ross writes that the decentralized no-collar workplace "'liberates' workers by banishing constraints on their creativity," the use of quotation marks makes explicit the sarcasm of the passage and, perhaps, reveals some of Ross's contempt for these duped workers.[106] Indeed, it seems to irritate Ross that companies like Razorfish are perceived by their employees as democratic, collaborative, and communal, without actually being worker owned and operated.

The second problem is that there seems to be something suspect in the very pursuit of leisure-at-work and, as I will argue shortly, the pleasure of gratification, enjoyment, or fun. As Ross writes at the end of No-Collar, "Paid employment that is most free from coercion often results in the deepest sacrifice of time and vitality" (again, this is the first problem).[107] He continues: "Nor does pleasure play fair. Gratification is no guarantee of justice, least of all in an economy that feeds on uncertainty and allocates rewards more unequally than it used to."[108] An analytical opposition is established here between pleasure and justice. Workers stand accused of prioritizing a "humane" workplace over a just workplace, with the implication that these workers might be a little too superficial or self-indulgent, at their own peril. This is also the subtext of a passage in which Ross describes corporate America's anxiety about changing workplace norms:

The informality of the no-collar workplace and work style was an obvious symbol of . . . nonstandard arrangements, and so the spread of
casual dress and like-minded liberties into Old Economy companies
was closely watched. The traditional managerial class resented the appeal of organizations where privileges could be enjoyed without serving due time in corporate ranks. But what they feared more was the
normalization of . . . Saturnalia [the ancient Roman festival in which
the roles of master and slave were reversed for a day] and the prospect
that the granting of autonomy would be taken for granted by employees who had no sense of rank-ordered etiquette or patience for proto-
col and who had an abiding infatuation with changing things to suit
themselves.[109]

Surely Ross is in favor of granting (true) autonomy to workers, though
he also seems ambivalent about this insofar as—according to his own
account—empowered workers have prioritized a "humane" workplace
over and above a "just" workplace, even though more than a few of
the employees interviewed seem to share Ross's political proclivities,
if in a vague or inchoate way. (Were his book written a decade later,
this interest in enjoyable work might have been ascribed to millennial
entitlement.)

This line of argument is strongly undergirded by a devaluing of forms
of leisure understood as irresponsible and self-indulgent, just as playbor
can only be understood as exploitative if it is seen as lacking in social
value, and by a corresponding valuing of responsible and collectively
oriented—that is to say social—relations. In Ross's framework, these
values are expressed through a juxtaposition of the "humane" and the
"just," where the "humane" is associated primarily with self-gratification
and the "just" is associated with the collective, the communal, the responsible, the social. If justice is concerned with fairness (as Ross suggests above), an investment in justice is predicated on a kind of social
awareness and management through which inequality can be amelio-
rated. Insofar as the humanization of the workplace (whether through
the transformation of work into something that scarcely resembles work,
or the introduction of elements of fun or play into the workplace) is understood as a self-serving project, it will always fall short of the social
ideal held up by critics.

## The Work Ethic

The argument that work and play are increasingly indistinguishable implicitly raises the question: What is the nature of the distinction between work and play such that this difference might collapse? In this section I offer an answer to this question—that is, that work can be distinguished from play by the presence of coercion—thereby calling into question the very notion of playbor.[110] Rejecting the notion that work and play are increasingly indistinguishable (on theoretical rather than empirical grounds), I propose instead that the anxiety about this collapse expresses an attachment to work as a symbolically important social institution. In other words, if the notion that play is transforming into work or work is transforming into play is unsettling to critics, this is because these transformations would threaten the integrity of work as a symbolic means of both establishing and exhibiting one's submission to the social. The source of this threat is precisely those forms of play/fun/amusement that are unsettling to critics insofar as they are understood as engendering antisocial hedonism. Keeping work and play separate is thus important for critics not because enforcing this distinction might protect workers/users from being exploited, but because it protects a sacred symbolic social milieu from contamination by the self-serving, individually oriented forces or drives that threaten to tear it asunder.

* * *

Hesmondhalgh's and Andrejevic's contributions to the playbor debate usefully identify coercion as a defining characteristic of work (though, for them, not all forms of work) insofar as one must work in order to have one's basic needs met; it is not one's employer but the political-economic system that forces workers to work. So while workers, if they are lucky, may choose a job from a host of alternative and presumably less desirable jobs, they generally do not have the choice to not work; they have "the liberty to work or to starve," as Herbert Marcuse sarcastically put it.[111] Whether or not work is individually pleasurable, it is thus structurally coerced. The converse is also true: when an activity is no longer structurally coerced, it no longer makes sense to understand it as work.

There is a simpler way to explain this, as I sometimes ask students in my Introductory Sociology class when examining work as a social institution: What would you rather be doing on any given Monday morning? Of course there are always students who insist that they, or more often their parents or guardians, love what they do and would do it for free, but even these students will often acknowledge the melancholy that accompanies the end of the weekend—insofar as weekends still exist—including for those who find their work pleasurable, fulfilling, or rewarding outside of monetary compensation. Furthermore, when work is the preferable option, this may say more about one's life outside work, which can be a greater source of struggle and strife, than it does about the desirability of work itself—a source of consternation for sociologists like Arlie Hochschild.[112]

A corollary to the idea that work is defined by coercion is that work is a function of the general (that is, aggregate) desirability of any given activity, or of particular organizations of that activity. So while it is possible to take pleasure in one's work, pleasure is not simply beside the point, as Andrejevic suggests; it is easier to find a job washing dishes than watching television. To reiterate, this is not simply a question of individual determination, such that one person's work might be another's pleasure; one can hunt as a hobby or for a living, as Thorstein Veblen long ago observed.[113] Rather, at an institutional level, work is organized around aggregate desirability in much the same way that Marx theorized exchange value as a measure of aggregate (in his terms, "socially necessary") labor time. This contextual element helps to explain how activities previously organized and waged as work can stop being work, and instead be performed for free by fans, enthusiasts, and do-gooders—a shift made possible (in part) by the extension of digital network technologies into everyday life. For example, video-game players identify bugs and contribute game modifications, consumers provide in-depth product reviews, journalism is increasingly the purview of "netizens," as are what used to be called "encyclopedias." In these cases, informal online participation replaces formal employment. In short, if enough people want to (and are able to) do something of their own accord, that is if they do not need to be coerced, then the activity in question does not need to be organized institutionally as work. The reverse is also true: activities once organized outside the purview of work can be institutionalized and formalized as work. For example, the displacement of manufacturing jobs by service jobs in the United States in the 1970s occasioned a formalization of certain capacities—"a good attitude and social skills" as Michael Hardt and Antonio Negri write—necessary for these new jobs: the work of customer service, telemarketing, and so on.[114] Work can thus be defined not by the production of value (insofar as leisure activities too can produce value), but by coercion as a function of the aggregate desirability of activities, in conjunction with their demand.

With this in mind, I would like to suggest that if online activities do not "feel, look, or smell like labor at all," as Scholz has written, perhaps it is because they are not labor after all.[115] This does not mean that critiques of playbor or leisure-at-work are not *about* work in another important way. As I will suggest, these critiques express an attachment to work as a symbolic object through which sociality is discursively established. To put it another way, what appears to be a concern about the exploitation and alienation of labor in analyses of playbor and leisure-at-work is also, and perhaps more fundamentally, a concern about a contraction of the work ethic, which itself is simply a means through which sociality can be established.

A few examples will help to illustrate how work serves as a symbolic object through which sociality is discursively established, and why a contraction of the work ethic would be so troubling to critics. In September of 2011, in the middle of his re-election campaign, Barack Obama—facing harsh criticism from opponents for failing to ameliorate the nation's ongoing recession—delivered a speech to a joint session of Congress outlining a piece of legislation (the American Jobs Act) that he promised would jumpstart a long-stagnant economy and reduce the unemployment rate. The speech was well-received though unexceptional, peppered with rhetoric celebrating the industrious but frustrated American worker:

> [Millions of Americans] have spent months looking for work. Others are doing their best just to scrape by—giving up nights out with the family to save on gas or make the mortgage; postponing retirement to send a kid to college. These men and women grew up with faith in an America where hard work and responsibility paid off. They believed in a country where everyone gets a fair shake and does their fair share—where if you stepped up, did your job, and were loyal to your company, that loyalty would be rewarded with a decent salary and good benefits; maybe a raise once in a

while. If you did the right thing, you could make it. Anybody could make it in America.[116]

In this passage, Obama conjures an archetypical American worker: industrious, disciplined, and fair-minded. This worker wants a "decent salary and good benefits," but also "maybe a raise once in a while." Why only "maybe" and "once in a while"? Because this worker is not greedy or demanding, but simply wants what is fair, what has been earned not only with work but with loyalty (to the company), sacrifice (for the family), and a general sense of responsibility.

A similar discursive formulation appears in Obama's 2013 State of the Union address, with a nod to the movement for same-sex marriage: "It is our unfinished task to restore the basic bargain that built this country— the idea that if you work hard and meet your responsibilities, you can get ahead, no matter where you come from, what you look like, or who you love."[117] In both these passages, work is imagined as a necessary though not-quite-sufficient condition for getting ahead; one must also be responsible and "[do] the right thing." In this way, the "bargain that built this country" is not simply a labor contract—so much work for so much pay—but also a social contract that solicits general obedience in the form of loyalty, tolerance (for those who "love" differently), responsibility, and accountability to the collective, which here takes the form of the nation. In short, it is not enough to work; one must also have a work ethic.

While it is easy to dismiss these kinds of political statements as vacuous and opportunistic, they reveal something important about the work ethic: that it is less about valuing work in and of itself than it is about valuing relations of responsibility, accountability, sacrifice, and general obedience. Work is simply the undesirable symbolic object through which one's selflessness can be discursively established. This is quite a feat, insofar as work is structurally coerced; it is not so easy to make an ethic of something that most people are forced to do to survive.

To establish this selflessness, one must first identify with the least gratifying aspects of work—the tedium, the exhaustion, the sacrifice— since it is suffering that both produces and evidences one's work ethic and, thereby, one's acquiescence to the community or collective. Work must be something one endures. Blue-collar labor has a lot to offer in this regard. One imagines backbreaking physical work, the drudgery of coal mines, farm fields, and assembly lines, what is sometimes called "honest work." White-collar labor, too, exacts its pound of flesh: years of schooling, the tedium of pushing papers, long hours staring at computer screens, intellectually taxing problem-solving, emotionally draining social contact, and stressful negotiation of time-sensitive issues.

Importantly, this identification with the least gratifying aspects of work need not be related whatsoever to one's valuing of actual work. Just as one can work tirelessly without a work ethic—for the money, for example—so too can one believe in the value of work without actually working. Rather than signaling the extent to which one values actual work, an identification with work sends a different message: that one is willing to suffer for unselfish reasons. Sacrifice is what makes an ethic of work; otherwise it is simply a way to make ends meet. Soldiers are noble; mercenaries are not.

Another brief example will help to illustrate this point. In my Introductory Sociology class, we spend a few weeks discussing economic inequality. In these discussions, students from wealthy families sometimes defend their wealth on the grounds that they (or rather their parents) have worked hard to get where they are. Judging by the reactions of other students in the class, the most compelling of these narratives entail long hours, arduous work, and social mobility; these are stories of suffering to get ahead. Many students, well-to-do or not, seem to have developed an allergy to the idea of getting something for nothing, an allergy made more acute by the financial crisis of 2007–2008 and, in response to this crisis, the rhetoric of Occupy Wall Street and the 1% (or 99%). If it is not we who have suffered, then we can be certain to find an other somewhere else who has suffered so that we might enjoy. Better, it seems, that it be us, at least in theory—a point to which I will return shortly. Through claiming to have suffered (or for their parents to have suffered) and thus to have earned their wealth, students defend their wealth by reference to a work ethic. Even though this wealth might suggest that work is a means rather than an end in itself, here wealth functions as a symbolic testament to having suffered.

While students from wealthy families can be quick to profess their work ethic in this way, many of them will also—when prodded gently—admit to some experience with various kinds of academic shirking (as will students from other socioeconomic backgrounds): working as little as they have to in order to secure desired grades, delaying work on assignments until the last minute, and avoiding forms of work that are particularly intellectually taxing. This might appear hypocritical: students say they value work, but in fact defer it whenever possible. However, unlike work in the cultural imaginary, actual work can be (and, I suspect, often is) undertaken not for its own sake, but for the promise of reward, financial or otherwise. As students well know, there need not be an ethic attached to work; it can also be a means to an end—money, or high grades in the hope of future money—rather than possessing value in and of itself. As Tim Kreider notes in the *New York Times*, "The Puritans turned work into a virtue, evidently forgetting that God invented it as a punishment."[118]

The wage thus complicates the work ethic: it is a reward for labor, but also a hindrance to displaying obedience, insofar as obedience entails selflessness and sacrifice, and the wage can facilitate—through the market—irresponsible forms of self-indulgent leisure. It is perhaps for this reason that Che Guevara argued that "labor should not be sold like merchandise but offered as a gift to the community," a point discussed in more detail in chapter 4.[119] To position himself on the righteous side of this thin line between reward and indulgence, Obama must admonish "those who prefer leisure over work, or prefer only the pleasures of riches and fame" (as he did in his inaugural address), even as he describes work as a "bargain" made to get ahead.[120] Or as Hillary Clinton put it in a campaign speech, "The people taking care of our children and our parents, they deserve a good wage, good benefits, and a secure retirement."[121] Presumably all workers "deserve" these things, but those who perform the work of "taking care" are highlighted—elevating this kind of work over and above other kinds of work. This is a paradox: selflessness cannot be motivated by a desire for money, but one deserves payment for having been selfless. In the words of Donna Summer: she works hard for the money so you better treat her right. Max Weber famously argued that this is why ascetic Protestants made for good capitalists; driven to work tirelessly to prove their elect status, yet forbidden from self-indulgence, they could only reinvest their earnings as capital.[122] In their attachment to work as a symbolic object through which the selflessness required for social bonds is established, critics of playbor and leisure-at-work are perhaps closer to this "spirit of capitalism" than they would care to admit.

## Antisocial Media

Just as it is possible to work without a work ethic, so too is it possible to indulge in various forms of leisure without making an ethic of it. (Surely some of the anxiety surrounding leisure comes from something more than a passing acquaintance with it.) And just as work serves as a symbolic object through which selflessness can be produced, so too do certain forms of leisure take on symbolic significance as manifestations of an unbound ego, of following one's desire without regard for the good of the community or collective. It is precisely these forms of leisure that are thought to threaten or undermine the formation of responsible social subjects.

While a variety of forms of leisure have historically provoked anxieties about the dissolution of the social—for example, recreational drug use and dancing—the pleasures of media occupy a particularly central place in anxieties about leisure as an antisocial, solipsistic endeavor.[123] These pleasures—those of being a reader, listener, viewer, player, and user—have long concerned scholars and critics, particularly insofar as these pleasures are thought to weaken or undermine valued identities, capacities, and forms of relationality by facilitating an escape from real world responsibilities, a suspension of reason in favor of emotion, and a giving over of oneself to fantasy and to the passivity of being entertained. To be clear, this is not to say that media are, in fact, antisocial, or that this perspective dominates media scholarship; myriad books and articles on fandom, the sociality of meaning, and participatory media, to take a few examples, have argued precisely the opposite.[124] It is simply to point out that there is something about the consumption of media that seems particularly to provoke anxiety about the dissolution of the social. In connecting the two lines of scholarship examined above to a history of anxiety about media as antisocial, my aim is not simply to say, "This again." Rather, contextualizing the anxiety that surrounds leisure-at-work and playbor as forms of exploitation within this broader pattern of anxiety will help to clarify the precise problems that particular forms of leisure present for critics. To this end, I will review briefly a few exemplary modern and contemporary texts of media criticism.

In his 2002 book, *Media Unlimited: How the Torrent of Images and Sounds Overwhelms Our Lives*, Todd Gitlin argues that we are "supersaturated" with media.[125] Drawing from the work of Georg Simmel, Gitlin argues that audiences are complicit in this saturation, immersing themselves in a torrent of media in order to stimulate "disposable feelings"—a substitute for the more substantial feelings that emerge from real interactions with real people in the real world. Following Simmel, Gitlin suggests that real feelings have become inconvenient, anathema to the instrumentality that characterizes everyday life in societies structured by money economies. The feelings stimulated by watching television, on the other hand, are easily provoked and just as easily put away, a kind of junk food for the soul.

Gitlin's argument is reminiscent of an earlier argument made by Neil Postman, who was famously concerned that television audiences were amusing themselves to death, or, rather, were being amused to death by television.[126] For Postman, this was not a question of bad practice, as if enlightened viewing practices might somehow resolve the issue; television was understood to be an inherently flawed medium, geared toward entertainment rather than reflection and leaving little room for serious thought. In fact, Postman argued, television audiences watch television precisely the way it "wants" to be watched: distractedly and with a fickle eye. No kind of programming could escape or inoculate itself to this logic, including political fare like presidential debates and educational shows like *Sesame Street*. As Gil Scott-Heron famously put it, the revolution will not—that is, could not—be televised.[127]

Back farther still, in the 1940s, Theodor Adorno similarly worried that "radio music" had a soporific and infantilizing effect on listeners: "Under the aegis of radio there has set in a retrogression of listening. In spite of and even because of the quantitative increase in musical delivery, the psychological effects of this listening are very much akin to those of the motion picture and sport spectatoritis which promotes a retrogressive and sometimes even infantile type of person."[128] For Adorno, this "retrogression" was effected through a synergy of form and content, producing listeners Adorno described as "regressed, arrested at the infantile stage," like children "with a sweet tooth in the candy store."[129] And like children, Adorno argued, these listeners "again and again and with stubborn malice . . . demand the one dish they have once been served."[130] Media technologies, in Adorno's formulation, are akin to a hypodermic needle or magic bullet—both these images have been used to describe the dynamics of transmission theorized by Adorno and like-minded critics—injecting passive audiences with ideology through both representational and aesthetic means.

Of course, the advent of media technologies also generates enthusiasm as well, typically before cynicism sets in. For example, the widespread adoption of digital network technologies in the 1990s was often seen as a cause for hope, occasioning renewed investment in a constellation of utopian desires—a rebirth of the public sphere, a restoration of the commons, a democratization of cultural production, and an erasure or liberation of identity markers tied to relations of oppression and exploitation—though also sometimes engendering a nascent dystopian sensibility that imagined these technologies as occasioning, if not directly causing, a destruction of mind, body, society, polis, and economy.[131] As backlash now seems invariably to follow initial periods of hype, so too has the utopianism made possible by the novelty of the Internet given way to this dystopian sensibility; as novelty fades, and the promise of new technologies is not brought to fruition, perceived patterns of actual use provide a foundation for dystopian anxieties. In fact, some of the most prominent contemporary dystopian media critics were once among the most ardent utopians.[132]

It is difficult to pinpoint when exactly the Internet lost its luster for critics, but a series of popular and well-received academic books published in the early 2010s seems to indicate the crossing of this threshold: Nicholas Carr's 2010 book, *The Shallows: What the Internet Is Doing to Our Brains*, Sherry Turkle's 2011 book, *Alone Together: Why We Expect More from Technology and Less from Each Other*, Jaron Lanier's 2011 book, *You Are Not a Gadget: A Manifesto*, and Evgeny Morozov's 2012 book, *The Net Delusion: The Dark Side of Internet Freedom*. Taken together these books provide a comprehensive sense of the dangers of living in the digital age, from the biological (Internet use atrophies neurological faculties, such as memory and attention), to the psychological (Internet use impedes emotional development), political/social

(Internet use weakens social bonds, the public sphere, and democracy), and economic (Internet use exacerbates exploitation, erodes the middle class, and undermines the stability of the economy).[133]

A similar sensibility has surfaced in popular culture. For example, in two separate commencement speeches, acclaimed novelists Jonathan Franzen (in 2011 at Kenyon College) and Jonathan Safran Foer (in 2013 at Middlebury College) both expressed concerns about the impoverishment of human relationships at the hands of digital network technologies. "I worry that the closer the world gets to our fingertips, the further it gets from our hearts," Safran Foer lamented.[134] Both writers advocated for a retreat from life online, and for "[putting] yourself in real relation to real people," as Franzen put it, rather than in relation to people online who serve as "diminished substitutes" as Safran Foer put it.[135] For what it's worth, Franzen also confessed to an infatuation with his Blackberry. Both speeches were adapted for publication in the *New York Times*, Franzen's under the title of "Liking Is for Cowards. Go for What Hurts," and Safran Foer's of "How Not to Be Alone."

The problem for critics is not simply that audiences mistake representations for the real, but that, like Postman's television audiences, they escape the real through their machinic attachments, as when Scholz laments that "time spent online means that we have less room in our lives to spend with friends—sitting together in the park or playing volleyball. Sometimes it seems that there is also less room for love, attention, and caring."[136] In these constructions, face-to-face interactions are conflated with valued forms of collective, communal relationality (as in the colloquial distinction between real life and life online). For example, Jodi Dean argues that when users experiment with identity online, they lack "the sense that an identity, once chosen, entails bonds of obligation."[137] She writes, "Rather than following norms"—what we should be doing— "we cycle through trends."[138] Dean takes particular issue with the "motions of clicking and linking," which

> do not produce symbolic identities; they are ways that I express myself— just like shopping, checking my friends' updates, or following tabloid news at TMZ.com. I may imagine others like me, a virtual local, but this local remains one of those like me, my link list or followers, those who fit my demographic profile, my user habits. I don't have to posit a collective

of others, others with whom I might need to cooperate or struggle, to whom I might be obliged, others who might place demands on me. The instant connection of networked association allows me to move on as soon as I am a little uncomfortable, a little put out.[139]

Dean worries that the subjectivities produced through online practices "may well be more accustomed to quick satisfaction and bits of enjoyment than to planning, discipline, sacrifice, and delay," virtues central to the "ground-level organizational work of building alternatives" to capitalism.[140] In conclusion, she writes, "In the circuits of communicative capitalism, convenience trumps commitment."[141]

Recalling the distinction between the social and the relational (established in the previous chapter), one can see that the problem for critics with particular forms of leisure or play online is not that they are anti-relational, but rather antisocial. In fact, the forms of leisure (online or not) targeted by critics are problematic precisely insofar as they are thought to engender devalued forms of relationality. For example, many of the pleasures of consumer capitalism are quintessentially relational, emerging from our contact with each other and with the world, but this contact is rarely thought to be obedient, responsible, sacrificial, disciplined, active, self-controlled, or autonomous; rather, it is considered to be docile, managed, intemperate, flighty, penetrated, and passive. For example, Geert Lovink laments that the social is "no longer the dangerous mix of politicized proletarians, of the frustrated, unemployed, and dirty clochards that hang out on the streets waiting for the next opportunity to revolt under whatever banner," but rather "a graph, a more or less random collection of contacts on your screen that blabber on and on— until you intervene and put your own statement out there."[142] Lovink writes that this transformation "troubles us theorists and critics who use empirical research to prove that people, despite all their outward behavior, remain firmly embedded in their traditional, local structures," like family, church, and neighborhood.[143] Together these statements make Lovink's empiricism seem less like a description or diagnosis than a directive: we must return to the social.

For these critics, if there is "commons" in "communication," it is not the glue that binds audiences or users together through webs of meaning, nor the shared cultural forms and traditions passed down and modified by new generations. On the contrary, it is a nightmare of promiscuous and irresponsible relationality, a delight in fleeting pleasures that can thwart even the most robust attempts to interpellate consumers or users into this or that subject position. This is why even consumers with the most critically defensible taste—the listener of National Public Radio, the reader of the *New York Times*, the fan of quality television—must still, at the end of the day, put down the book, turn off the television, and put away the phone/laptop/computer; the (queer) medium is the message.

Neither is the problem simply a downward slide from civic-mindedness to hedonism, but more precisely what is imagined to be the passive character of this hedonism. For critics like Postman and Gitlin, the television is a "boob tube" not only in the sense of "idiot box," but in the way it infantilizes the viewer who nurses at it. Academic treatments of culture and media have long valued the active over the passive. This is true not only of Frankfurt School critiques of consumption as terminally passive, but of the hermeneutic response to these concerns (popular, especially, in cultural and media studies), which emphasizes the sociality of meaning, so that what would appear to be a catatonic couch potato, for example, might be revealed as actively engaged in a process of contesting dominant ideologies.[144] This valuing of the active is also present in more conventional approaches taken by scholars in sociology and communications, as in Uses and Gratifications Theory.[145] In short, empirical disagreements about the extent to which audiences/users are active or passive mask underlying political and ethical lines of affinity; critics may understand differently, but they desire similarly.[146]

This common valuing of the active over the passive brings to mind Bersani's insight (via Foucault) that "the only 'honorable' sexual behavior 'consists in being active, in dominating, in penetrating, and in thereby exercising one's authority."[147] "To be penetrated is to abdicate power," Bersani writes, and it is precisely through this abdication that pleasure is produced.[148] The penetrated bottom, passive in discourse if not in practice, is intolerable for the same reason that the passive audience hypodermically injected with dominant ideology is intolerable. In both cases, there is a perceived refusal not just to assume power but, beneath this, to identify with the collective in such a way that would make this assumption desirable.[149] This, perhaps, is critics' underlying problem with passivity: it involves a lack of interest in or disinclination toward responsible subjectivity, what Bersani calls "a revolutionary inaptitude . . . for sociality as it is known."[150]

Even now that passive audiences have been transformed into potentially active users in the eyes of those concerned, trouble remains. If anything, the squandering of the active potential of digital network technologies is even more disappointing for critics than the inevitable passivity of earlier forms of media. If the Internet is an especially frustrating medium for critics like those examined above, this is because Internet users are thought to come close to being active, unlike television audiences, which, for critics like Postman, might be rescued only by turning off their television sets and crawling back to reality like Plato's bleary-eyed puppet-show dupes emerging from the cave. It is from this position of frustration, for example, that Evgeny Morozov implores, "We need to find ways to supplant our promotion of a freer Internet with strategies that can engage people in political and social life. Here we should talk to both heavy consumers of cat videos and those who follow anthropology blogs. Otherwise, we may end up with an army of people who are free to connect, but all they want to connect to is potential lovers, pornography, and celebrity gossip."[151]

The opposition in these texts to forms of leisure understood as self-indulgent or as otherwise irresponsible helps to explain why so many on the Left seem attached to the notion that pleasure must always come at the cost of suffering, whether it is the suffering of those who indulge, or of those whose labor makes possible others' indulgence. While this will to identify complicity appears to be motivated by an altruistic concern for the immiserated other, my argument here suggests that it more fundamentally expresses a deep-seated discomfort with forms of pleasure that release us from social bonds, from our responsibility to this other, our attachment to which may be less selflessly motivated than critics would like to believe. Underlying these accounts is a notion that pleasure must be earned through suffering or some other form of acquiescence to the social in order to be deserved. In this way, pleasure is imagined as a sort of zero-sum game, with no more pleasure allowed than suffering has made possible; it seems scarcely possible to imagine a world in which pleasure might be both abundant and free.

The notion that the boundary between work and leisure is shifting or disappearing, and more precisely that activities that appear to be leisure are actually work or work-like in significant ways, is thus important to scholars not simply because users' or workers' value is appropriated, nor because they have been alienated from their own productivity, but because of an underlying attachment to work as a symbolic object through which responsible sociality is established, asserted, and maintained. This attachment to work is expressed and furthered through the anxiety that surrounds the dissolution of valued forms of relationality at the hands of particular forms of leisure. Even if debates about the exploitation of leisure have already exhausted themselves, as Geert Lovink has suggested, these debates still offer a window into the normative underpinnings and aims of critical political-economic Internet and media scholarship. One might say, in fact, that these arguments aim to redescribe leisure as work (rather than simply empirically reporting this transformation) in order to solicit readers to assert forms of self-control, self-governance, and other-directedness that critics fear they will avoid or escape precisely through targeted forms of leisure.

The anxiety surrounding the transformation of leisure into work thus aims to reestablish a boundary, not between labor and leisure, but between responsible and irresponsible forms of relationality, produced discursively through desirable and undesirable symbolic objects or activities (particular forms of leisure and work, respectively). In expressing anxiety about the collapse of work and leisure, scholars thus create discursive space for the production of responsible subjects. Incidentally, this attachment to work as a discursive vehicle for collective relationality forecloses other frameworks by which one might make sense of these phenomena. For example, users might be understood not as working, but as providing raw material for the work performed by algorithms, or by statisticians, as Goran Bolin has suggested.[152]

The irony in all of this is that this attachment to work seems to take the form of its opposite: an identification with overworked, underpaid, exploited laborers, or users who have been taken advantage of, who have been alienated from their own productivity, and who will never collect the economic value produced through their own participation. But critics of playbor or leisure-at-work can never be content with workers' or users' desires for more—more money, more fun, more stuff—insofar as these demands risk overindulging the greedy ego, threatening valued forms of collective relationality. Thus, it is not surprising that arguments for the reduction of work in favor of work/life balance often end up valuing other social forms, like spending more time with the family or participating in the community. This is also to recognize that it is possible to advocate for less work and, at the same time, value collective relationality, which can be established and maintained through other institutional alignments.

From a queer perspective, and particularly following the antisocial thesis, the refusal or disinclination to work falls short of its antisocial potential if work will simply be replaced by other institutions through which valued forms of relationality are made normative. Referencing the slogan of the eight-hour movement—"eight hours labor, eight hours rest, eight hours for what we will"—Kathi Weeks writes, "Rather than, for example, appealing primarily to norms of family responsibility, this formulation suggests that a movement for shorter hours should be animated not only by the call of duty but also by the prospect of pleasure."[153] Is it possible to go one step farther here and ask whether such a "movement" need be animated by a call of duty at all, which, in the end, is just another form of work? Must one form of work always be traded for another, or might we forgo this attachment to work altogether?

# 3

## Automating

Idle hands are the devil's workshop.
—Proverbs 16:27

Recall the future as imagined in the 1960s and 1980s cartoon *The Jetsons*. Machines cook and clean; there is little need to lift even a toothbrush. People rarely walk from place to place; instead, they're shuffled around on moving sidewalks. In one episode ("The Vacation") George Jetson sits at a giant computer pushing a single button and says, "Boy this job is a killer. I spend an hour a day, two days a week working my fingers to the bone." In twentieth-century U.S. popular culture, the promise of technology was often less work, more leisure, and more stuff—in a word, abundance.

It is not yet 2062 (when *The Jetsons* is imagined to take place), but the automation of labor in contemporary society has, in some ways, exceeded the cartoon's fantastical projections. Rather than signaling abundance, however, automation is widely understood as engendering scarcity, and in particular a scarcity of jobs, as illustrated by the ongoing publication of news stories about "technological unemployment" beginning in 2011. A sampling:

"Will Robots Steal Your Job?" (A five part series, *Slate*, September 2011)
"Skilled Work without the Worker" (*New York Times*, August 2012)
"Robots and Robber Barons" (*New York Times*, December 2012)
"Better than Human: Why Robots Will—and Must—Take Our Jobs" (*Wired*, December 2012)
"Robots Are Already Replacing Us" (*Wired*, December 2012)
"The Robots Are Coming" (*Washington Post*, January 2013)
"Recession, Tech Kill Middle-Class Jobs," "Practically Human: Can Smart Machines Do Your Job?" and "Will Smart Machines Create a World without Work?" (Associated Press, January 2013)

"What Jobs Will the Robots Take?" (*Atlantic*, January 2014)

"March of the Machines" (*60 Minutes*, January 2014)

"Coming to an Office Near You" (*Economist*, January 2014)

"10 Jobs Replaced by Machines" (*Mashable*, January 2014)

"The Robots Are Coming. Will They Bring Wealth or a Divided Society?" (*Guardian*, January 2014)

"9 Robots that Are Stealing Our Jobs" (*Business Insider*, February 2014)

"Robots Will Replace Fast-Food Workers" (CNN, May 2014)

"Robot Doctors, Online Lawyers and Automated Architects: The Future of the Professions?" (*Guardian*, June 2014)

"Robots Work Their Way into Small Factories" (*Wall Street Journal*, September 2014)

"Can a Computer Replace Your Doctor?" (*New York Times*, September 2014)

"A World without Work" (*Atlantic*, July/August 2015)

"Will Machines Eventually Take Every Job?" (BBC, August 2015)

"Intelligent Machines: The Jobs Robots Will Steal First" (BBC, September 2015)

"Smart Robots Could Soon Steal Your Job" (CNN, January 2016)

"Yes the Robots Will Steal Our Jobs. And That's Fine." (*Washington Post*, February 2016)

"Are Robots Going to Steal Your Job? Probably" (*Guardian*, April 2016)

"Automation and Anxiety" (*Economist*, June 2016)

"A Robot May Be Training to Do Your Job. Don't Panic" (*New York Times*, September 2016)

As *60 Minutes* succinctly summarizes, "Instead of serving us, we find them competing for our jobs."[1]

How should we understand the anxiety that "robots" will "steal [our] jobs," and—beneath this—the notion that the disappearance of jobs is cause for alarm? One might expect the prospect of this disappearance to produce precisely the opposite reaction; rather than being a cause for alarm, the elimination of work might be a cause for celebration. After all, popular culture is thick with disdain for work. It suffuses our advertisements, song lyrics, and television shows. What happened to the fantasy expressed in *The Jetsons* of working an hour a day, two days a week?

Perhaps it goes without saying that it is not jobs that people want, but money; we have simply resigned ourselves to the fact that we must sing for our supper. Certainly much of the recent attention paid to technological unemployment—a term coined by economist John Maynard Keynes to describe unemployment caused by automation—treats this phenomenon as an economic problem.[2] For example, in one of the most prominent books to examine contemporary technological unemployment, *The Second Machine Age: Work, Progress, and Prosperity in a Time of Brilliant Technologies*, Erik Brynjolfsson and Andrew McAfee observe that while the gross domestic product (GDP) has increased over time, the relative share of the GDP allocated to labor has decreased. They offer two explanations for this trend: higher unemployment and declining wages for the employed.[3] They suggest that digital and network technologies are partially to blame, insofar as these increase productivity mostly to the benefit of those who own physical capital—the machines that produce wealth. This seems like a reasonable and uncontroversial explanation: fewer jobs and lower wages would certainly decrease labor's share of the GDP. Furthermore, these two trends seem complementary, particularly when tied to a narrative of technological unemployment. If technology is displacing laborers, then it stands to reason that there are fewer jobs, and that those who are working are working for less money, owing to the sizable standing reserve of unemployed laborers and to the ever-present threat of automation. As Brynjolfsson and McAfee conclude, "If digital technologies create cheap substitutes for labor, then it's not a good time to be a laborer."[4]

Contesting this economic framing, this chapter argues that for critics, concerns about automation and technological unemployment are about more than money, whether this money goes to the poor and unemployed (for whom we ought to care, if we are being ethical, and/or whose increased purchasing power will ultimately bolster the economy) or lands in our own self-interested hands. Or rather, the call for more jobs is not simply a call for more money for more people. If this is the case, as the chapter proposes, then the anxiety surrounding automation and technological employment should not be engaged at face value—for example, by debating the extent to which technology actually poses a threat to labor, and what should be done about this threat—but rather needs to be interpreted, which is also to say that this anxiety does not name or perhaps even know its own ends.

To begin to unpack the anxiety surrounding technological unemployment, we might ask: Apart from the wage, what do those anxious about automation and technological employment believe that we would have to give up if we were to stop working? One can learn a lot about a sacred object when its owner fears that it will be transformed or taken away. One can also learn a lot about its owner, and about the relation of owner to object. What are critics really trying to hold on to when they make the case for more jobs?

The previous chapter illustrated work's primacy as a symbolic object, particularly in terms of producing responsible subjects by way of the work ethic. To put it another way, work is social not only literally (insofar as work is often collaborative and interdependent), but symbolically; because work entails sacrifice/suffering, an identification with work allows a nascent social subject to exhibit the selflessness required of collective relationality. It follows from that argument that anxieties surrounding automation and technological unemployment are motivated not only by a desire to reduce unemployment, but perhaps more significantly by a desire to preserve forms of collective relationality to which the institution of work has been symbolically attached. In other words, automation, like leisure, is unsettling insofar as it threatens to undermine an institution through which responsible subjects are formed. But there is an additional and essential way that anxieties about automation and technological unemployment participate in the production of responsible subjects, which resides in the "robots" that are thought to be the agent of this transformation.

It is significant that technological development has been singled out as the cause for unemployment by scholars and journalists concerned about automation, while a number of other threats to employment have evaded scrutiny; once again, it is this inconsistency that suggests that scholars' and journalists' concerns (in this case, about automation and technological unemployment) should not be taken at face value. For example, critics might target volunteerism, crafting, and choring (mowing the lawn, sewing and knitting, coaching children's athletic teams, and so on)—activities that displace landscapers, garment workers, and other laborers. Walking deprives taxi drivers of a fare, auto manufactures of a sale, and gasoline companies of a customer. If these various kinds of voluntary activity could be curtailed, the supply of jobs would increase relative to the supply of labor, and wages would go up, insofar as flooded labor markets drive wages down.

To take another example, the satiation of consumer desire might also be a cause for concern. If the goal were simply to create more jobs, one might advocate for stimulating consumers' desire for more products and services. Consumer capitalism has repeatedly shown that human desire affixes easily to new objects, the production of which could stimulate employment. Tellingly, those concerned about automation and technological unemployment rarely take up this cause. It seems that as consumers, we are supposed to want less, not more, whether to forestall environmental catastrophe or simply to establish our moral or ethical worth as temperate, self-disciplined, and not greedy. Of the authors considered below, only Martin Ford argues that consumerism (not labor) is the true foundation of the economy and needs to be protected as such.[5]

Finally, one might consider competition for jobs within labor markets as cause for concern. Laborers compete not only against "robots" and volunteers, but against each other, particularly in a globalized economy in which many jobs have been outsourced overseas. Historically, labor markets have been managed in two ways to reduce this competition: (1) through exclusions from the labor market, as by trade guilds (often along the lines of nationality, race/ethnicity, and sex/gender), and (2) through population management, whether through state policy or informally through social means.[6] Interestingly, the "desires" of capital line up rather neatly with progressive ethics on this point: capital wants open labor markets (to drive down wages), as do progressives—not to drive down wages, but in solidarity with structurally excluded workers and in opposition to conservatives (like Pat Buchanan, for example) whose xenophobia is seemingly deployed in the interest of the "American worker." This creates a bind for progressives, who must find other ways to exclude potential workers from the labor market, for example, through mandatory retirement policies or raising minimum age requirements.

Furthermore, even if one were to agree that technology represents a singular threat to the institution of work, one could still easily argue that a reduction of work need not necessitate a reduction of wages. On the contrary, as Jonathan Cutler has suggested, it is possible for workers both to work less and earn more.[7] In particular, Cutler argues that when the supply of labor goes down, the cost of labor goes up; when workers sell less of their labor in the market, they can demand higher wages for it. What workers need, he suggests, is not more jobs, but rather to be lazier and more entitled, to indulge their inertia, reducing the supply of labor relative to demand and thereby driving wages up. Following this argument, efforts to mitigate technological unemployment are misguided—that is, if their true goal is to secure workers' livelihoods. Rather than focusing on technology, one might focus instead on the cultural standards surrounding work: how much it takes to move workers to labor, and the conditions under which they will agree to labor.

Interestingly, Brynjolfsson and McAfee articulate this very point, though only in relation to the earnings of "superstars," noting that "cultural barriers to very large pay packages have fallen. CEOs, financial executives, actors, and professional athletes may be more willing to demand seven- or even eight-figure compensation deals."[8] What Brynjolfsson and McAfee fail to recognize is that what is true of these superstars could also be true of ordinary workers, though this recognition would undermine what seems to be the primary purpose of highlighting superstar earnings—namely, to admonish their greed. But, as Cutler's argument suggests, workers have much to learn from the greed of superstars.

Why, then, has technology been singled out as a threatening agent? For one, unlike other plausible causes of unemployment and underemployment, automation is not understood as supporting cherished values. On the contrary, automation, like consumerism and particular forms of leisure, is understood as a threat to values that can be asserted only through institutions (like work) that require subjugation. In the case of automation, it is the very institution of work that is threatened by automation's promise to put an end to work.

More specifically, this chapter will argue that technology has been singled out as a threatening agent because of its neutral, docile character—or rather, what is understood to be its neutral, docile character. This apparent docility figures in two central ways. As examined in the first section of the chapter, it makes it possible to imagine technology as that which channels and serves the will of its master—typically identified as those who control the means of production—with pernicious social and political-economic side effects that can be mitigated only through state intervention. In this account, the state and social institutions are summoned to act both in the interest of workers to promote job growth and as a buffer against the vicissitudes of the market or, in a Marxist variation, against the power and control of capitalists. Here, technology—and the capitalists it serves—is mobilized discursively as a force against which workers need to be protected, while the state serves as a discursive proxy for the social.

The second section of the chapter considers the other primary way in which the apparent docility of technology figures: in accounts of technological unemployment, the production of technology as a neutral, docile object provides a discursive foil against which an active, self-governing laboring subject can be asserted. This is also to say that the reason that technology is singled out as a threat to (human) labor is that it is understood as imperiling distinctly human qualities of mind, which are valued, in part, for their discursive proximity to collective governance. Drawing from social and political-economic theory, including work by Marx and Engels, George Caffentzis, and Sara Ahmed, the section proposes that this is why contemporary anxieties surrounding technological unemployment primarily target the automation of cognitive, professional, white-collar labor. If to be human is (a) to be engaged in this kind of labor, and (b) to have the capacity to resist working, as well as the capacity to engage in collective governance, then it should not be surprising that technology registers as a threat to this rather particular notion of the human and of the social that it serves. This section thus argues that technology is not so much a threat against which workers need to be protected; rather, it serves as a symbolic object against which workers can be discursively distinguished as collectively oriented social subjects.

In this way, the chapter argues that the anxiety surrounding technological unemployment serves as a call to two related forms of governance: state governance and collective governance. According to the "logic" of this anxiety, laborers need the state and social institutions to protect them from capital, and/or they need to become self-governing social subjects. In the first case, the state is charged with maintaining the social. In the second case, this is charged to laborers themselves. In both cases, the call to governance works to establish and preserve valued forms of relationality built on bonds of responsibility, whether to others or over oneself and one's labor.

## Calling the State

In the first call to governance, the figure of the robot—or, in a variation of this argument, the owner of the robot—is conjured as a threat from which workers need to be protected by an overseeing third party, typically the state. To put it another way, anxiety about technological unemployment summons the state to rescue the social from the threat of technology. This anxiety sometimes manifests in predictions about what will happen without state intervention. These imagined futures then provide a rationale for state intervention. For example, in his 2009 book *The Lights in the Tunnel: Automation, Accelerating Technology and the Economy of the Future* Martin Ford warns,

> If we do not have a strategy—and specific policies—in place to deal with this issue before its full impact arrives, the outcome will be decidedly negative. As the trend toward systemic job loss increases, it is quite easy to foresee a number of possible ramifications. I have already mentioned the likelihood of a drop in college enrollment and a migration toward safer trade jobs. Another trend that will surely occur as recognition sets in will be a general "war on technology." Workers in virtually every occupation—even many of those who themselves work in technical fields—will desperately, and quite understandably, attempt to protect their livelihoods. We can expect substantial pressure on government to somehow restrict technological progress and job automation. It is possible that there will be a significant, last-ditch resurgence in the power of organized labor. Workers in jobs and industries that are not now organized will very possibly turn to unions in an attempt to exert some power over their own futures. The result is likely to be somewhat slowed technical progress, work stoppages, and significant economic and social disruptions.[9]

While Ford concedes that his predictions may not be entirely correct, he finds their probability sufficient cause to develop a mitigating plan; a risky future warrants a cautious present.[10] Here, anxiety surrounding automation and technological unemployment takes the form of a threat: have a plan or else face the possibility of social unrest and upheaval. It is not simply the future of the economy that is at stake here, but the future of the social. The "we" Ford continually employs in this passage refers not only to the social "we," but to the state, which is charged with protecting the social. In fact, Ford argues, it is only the state that has the power to intervene and stave off catastrophe.[11] Ford, it should be noted, is not alone in this view.[12]

The state, in this account, is solicited to take various kinds of action. For one, it is called upon to foster the development of new kinds of jobs that are safe from automation, if only in the short term, and to help prepare workers for these jobs. Education is often thought to be central to this process, insofar as workers need to be prepared for new jobs through retraining.[13] In addition to advocating for investment in and reform of education, Brynjolfsson and McAfee propose a number of other state initiatives to encourage job growth, including streamlining regulation in order to foster entrepreneurship; developing services and databases that evaluate potential employees and match them with jobs; increasing government funding for scientific research; balancing intellectual property rights to encourage innovation; increasing government investment in infrastructure to support manufacturing; and liberalizing immigration policies.[14] Again, these initiatives are less present-oriented than they are future-oriented. Or rather, present intervention is justified by the prospect of a future of mass unemployment, a disintegrating economy, and a fraying social fabric.

The state is called upon not only to intervene preemptively, but to mitigate the negative effects of technological unemployment—particularly economic inequality—in the present. As Brynjolfsson and McAfee argue, the economic gains of automation are not evenly dispersed among the population but rather concentrated in relatively few hands.[15] Jaron Lanier has termed this "winner-take-all capitalism," in which those who own and control fixed capital—the most powerful computers, in this case—are the only ones to profit.[16] To support this argument, Lanier contrasts Kodak, which once employed 140,000 people and was worth $28 billion, with Instagram, which employed 13 people at the time of its sale to Facebook for one billion dollars. "Where did all those jobs disappear to? And what happened to the wealth that those middle-class jobs created?" Lanier asks.[17] In a line of thinking familiar from the previous chapter, he reasons that these jobs are now performed by Instagram's users for free, the wealth from which is collected and enjoyed by Instagram's few actual employees.

In a variation of this argument, some scholars, journalists, and pundits contend that concerns about technological employment distract from the real, underlying problem: a struggle between labor and capital, rather than labor and technology. The real problem, in this view, is not unemployment—technological or otherwise—but rather systemically produced economic inequality. For example, Alex Payne suggests that critics are not really concerned about automation; they are concerned about the extreme concentration of wealth and power. Workers do not want to slow or stop automation, but rather to gain and exercise forms of leverage that would allow them to better control the conditions of their work.[18] Tellingly, for Payne, this is not simply about fairly distributing the wealth produced through automation—which might be done through the state—but about "self-determination," which is to say workers' transformation into self-governing subjects, a point explored in depth in the next section.[19]

If the problem with automation is that it either directly or indirectly produces a wildly uneven distribution of wealth, what is the solution? While Lanier stops short of calling for state intervention—an expression, perhaps, of his Silicon Valley libertarianism—many of those who understand technological unemployment as intimately linked to economic inequality advocate strongly for state intervention, which becomes necessary either because corporate executives refuse to share profits with workers, preferring instead to remunerate themselves extravagantly, as Lynn Stuart Parramore suggests, or because there are no workers left to pay, as Lanier suggests.[20] One form this proposed state intervention takes is taxation. For example, to mitigate the short-term effects of technological unemployment, Ford proposes a gross margin tax and progressive corporate tax deductions related to wage and salary expenses. His long-term suggestions include recapturing wages from automated jobs through additional taxes: business taxes, capital gains taxes, progressive income taxes, and consumption taxes. Brynjolfsson and McAfee similarly advocate for taxation, particularly a negative income tax and reduced taxes on both workers and employers, as a partial solution to the economic inequality caused by automation. This emphasis on taxation does not simply lend itself to state intervention, but requires it insofar as the state is posited as the only legitimate regulation and taxation authority.[21]

For some of these scholars, part of the problem with rising economic inequality is that laborers are also consumers; unemployment not only impedes workers' ability to provide for themselves and those they financially support, but also undermines the overall health and stability of the economy, which rests on mass consumption. As Ford argues, purchasing power depends on employment; with mass unemployment or under-employment, the market for goods and services stagnates.[22] Once this happens, disaster will ensue according to Brynjolfsson and McAfee, with weak demand for goods and services leading to more unemployment and a general deterioration of working conditions leading to even less demand for goods and services, and so on in a downward spiral; they term this a "failure mode of capitalism."[23] It is for this reason that Ford prioritizes the preservation of the market; insofar as mass consumption is central to economic health and stability, it must be protected.

For Ford, one form this protection must take is a government-issued income, an expense to be paid for through the forms of taxation he lists. However, we should not expect to get this money for free; Ford's plan calls for people to paid insofar as they participate in incentivized activities, the social value of which justifies remuneration: education, including continuing education and even just simple reading ("a more educated population has many benefits to society, including a lower crime rate, greater civic participation, a more informed electorate and a more flourishing cultural environment"); community and civic partici-pation; journalism; and the reduction of negative externalities such as environmental pollution.[24] This proposal makes clear that it is not only people's economic livelihoods that are at stake, along with the economy that secures these and is secured by these, but also the health and stabil-ity of the social. Insofar as work is a primary mechanism through which the social is maintained, the possibility of an end to work threatens to tear the social asunder. The state is thus solicited to act as an arm of the social, securing its future.[25]

This point is also made clear in Payne's public letter to libertarian tech investor Marc Andreessen. Payne writes: "Hopefully we can agree that there are many more meaningful quality of life improvements technol-ogy has yet to deliver on before we can start brainstorming the 'luxury goods markets' of the future. Meanwhile, we don't need to wait until a hypercapitalist techno-utopia emerges to do right by our struggling neighbors. We could pay for universal health care, higher education, and a basic income tomorrow. Instead, you're kicking the can down the road and hoping the can will turn into a robot with a market solution."[26] Here Payne contrasts the pursuit of luxury goods to "meaningful quality of life improvements"—improvements not for Payne, it seems, but rather for "our struggling neighbors."[27] The accusation is clear: tech investors have skewed priorities, valuing luxury—ostensibly for their own pleasure—over and above care for others. The "we" used liberally here seems to identify Payne not only with his readers but also with the state, which is summoned to pay on behalf of taxpayers for the meaningful improvements Payne lists: health care, education, and income.

To be certain, not all scholars agree that state intervention is necessary to rescue the social; some imagine that a solution is possible through more informal means. Lanier, for example, emphasizes the importance of the social contract, which is what provides the mandate for governance in the first place. This is clearly illustrated in the way he narrates labor history, reimagining the wage relation as a kind of charity offered by business owners to preserve social and economic well-being, rather than as a site of struggle between labor and capital:

> People often say, well, in Rochester, N.Y. . . . they had a buggy whip factory that closed down with the advent of the automobile. The thing is, it's a lot easier to deal with a car than to deal with horses. . . . And so you could make the argument that a transition to cars should create a world where drivers don't get paid, because, after all, it's fun to drive. And it is. And they're magical.
>
> We kind of made a bargain, a social contract, in the 20th century that even if jobs were pleasant people could still get paid for them. Because otherwise we would have had a massive unemployment. And so to my mind, the right question to ask is, why are we abandoning that bargain that worked so well?[28]

Lanier then elaborates:

> Of course jobs become obsolete. But the only reason that new jobs were created was because there was a social contract in which a more pleasant, less boring job was still considered a job that you could be paid for. That's the only reason it worked. If we decided that driving was such an easy thing [compared to] dealing with horses that no one should be paid for it, then there wouldn't be all of those people being paid to be Teamsters or to drive cabs. It was a decision that it was OK to have jobs that weren't terrible.[29]

According to Lanier, the impetus to pay workers for less unpleasant work is to keep unemployment down and preserve the middle class and, thereby, democracy. In contrast to arguments that blame greedy capitalists for economic inequality, Lanier recasts conflict as contract; mutually exclusive interests are reimagined as mutually beneficial. In this contract, capital not only offers wages for tasks that are really too easy, fun, or pleasant to be waged, but also maintains the position of the middle class and thereby secures the greater social good. The middle class, in turn, ought to be grateful; as Lanier asks, "Why should you even be paid to do anything?"[30] The social contract forged between capital and labor is thus thought to be grounded in compromise, sacrifice, and responsibility, with each side tempering its interests for the greater good. Insofar as work is central to this social contract, treating non-work as if it were work becomes a way to ensure that this kind of social relation continues. It is on these grounds that Lanier takes issue with companies like Google that offer information for free and generate revenue through advertising. This is a problem for Lanier not because these companies are greedy, but because they renege on that social contract.

Lanier's provocative solution to the problem of technological unemployment is not state intervention but micropayments given to ordinary users when the data they produce is used by a company. While Lanier does not suggest that companies like Google are exploitative, and while he shies away from characterizing Internet use as playbor, he nevertheless ends up with a similar politics as some critics of playbor, advocating for a redistribution of wealth, through for different reasons: the creation of a strong middle class, which he sees as central to the democratic project.[31]

Here we might question whether democracy is the only end of Lanier's argument—the thing he truly wants—or whether the social contract that delivers democracy via the production of a middle class is not equally valued. In fact, we might consider democracy and the middle class it supposedly represents as a kind of discursive object that stands in for the social contract, insofar as the middle class is not a "natural" class but is produced through social convention and state regulation, for example academic tenure, taxi medallions, cosmetology licenses, and pensions, as Lanier writes, as well as things like the micropayments for which he advocates.[32] To put it another way, the middle class and democracy function as already valued symbolic objects that legitimate the imposition of a social contract. Here the message is: if you care about the preservation of the middle class and democracy, then relations of sacrifice and responsibility must be maintained.

Whether the state is called upon to preserve the social from the market, or whether (as in Lanier's argument) market players are called upon to preserve the social from what appear to be the market's misanthropic, antisocial tendencies (explored in more detail in chapter 4), the effect is the same: a call to govern the social. The anxiety surrounding automation and technological unemployment thus produces a mandate for governance, which becomes necessary not only to prevent social and political-economic upheaval, but also to preserve valued relational bonds; as Payne puts it, governance becomes necessary to "do right by our struggling neighbors."[33]

## The Will to Work

In the second call to governance, technological unemployment is not a problem to be solved. Rather, the anxiety surrounding technological unemployment is itself a solution to an underlying problem: the loss (or failure) of a self-governing, laboring subject who identifies with valued forms of relationality. In projecting anxiety onto a technological object that has been differentiated from a human subject along the axis of willful, collective governance—robots are docile, but workers can self-govern—anxiety about automation creates discursive space for the formation of that very same self-governing subject.

To begin to explain how and why this anxious projection occurs, it is important to note that labor has been subject to automation for centuries. Furthermore, scholars and journalists writing about technological unemployment are clearly aware of this. As Lanier notes, the advent of the automobile industry rendered obsolete the manufacture of buggy whips.[34] Lamplighters were replaced by electric lights, and ice delivery by refrigeration. Neither is the anxiety surrounding technological unemployment novel; scholars and journalists are clearly aware of this as well. In 1930, for example, John Maynard Keynes was concerned about the automation of agricultural labor.[35] Those critical of the notion of technological employment often refer back even farther to the Luddites (followers of Ned Ludd/Ludlam, an eighteenth-century weaver who destroyed a weaving loom, apparently in protest of the automation of weaving). One might go farther back still, to scribes' protest of the printing press in the fifteenth century.[36]

Critics skeptical of the relation of automation to unemployment have been quick to point out the error of these earlier concerns, which consistently underestimated the creation of jobs following the adoption of new technologies. Not only are new jobs eventually created, critics contend, but technologies once thought to threaten labor are often later embraced precisely as the cure for labor's woes. In the years immediately following the Great Recession in the United States, for example, there was significant nostalgia for an industrial economy, as expressed in the romanticization of the manufacture of physical goods, blue-collar labor (even when this labor is at the helm of a machine, assigned a single, routine task on an assembly line), and blue-collar settings (the factory, the small business, the workshop).[37] In this context, certain kinds of machines, technologies, and tools no longer seem to pose a threat, but rather are understood to be the solution to the problems of a post-industrial economy.

While the cyclical and apparently misguided nature of concerns about automation might suggest to those anxious about automation and technological unemployment that these concerns are bogus, scholars and journalists have argued that this time will be different. What makes this time different, according to those concerned, is that there remain few kinds of labor that cannot be automated and that might otherwise provide a refuge for displaced laborers in the future.

To this point, those skeptical about technological unemployment respond that unemployment resulting from automation is less severe than proponents contend, that this time is not in fact different, and that we simply need to be patient for new kinds of jobs to emerge as they have in the past, as well as promoting the kinds of retraining that will be necessary to prepare workers for these new jobs. For example, Lynn Stuart Parramore argues that if technological unemployment were real, the advent of personal computing in the 1990s should have ushered in massive unemployment, but in fact the reverse occurred.[38] Citing a *Wall Street Journal* survey of economists, Parramore suggests that jobs will eventually return, "whether or not a Roomba is vacuuming the floor."[39] In a segment on automation for PBS, Vivek Wadhwa points out that the app economy—a job sector that could not have been anticipated a decade ago—now employs 500,000 people in the United States.[40] Rather than facing the obsolescence of labor, critics contend, we are simply facing its transformation.

This is not to say that jobs aren't continually rendered obsolete by technological development. In fact, as a 2014 Pew survey concludes, most experts agree that automation does in fact displace workers and that machine learning and robotics will indeed have a wide-ranging impact on employment and the economy, particularly in terms of job displacement.[41] However, critics argue that it is difficult to anticipate the new jobs that will be created in the wake of technological change, as Wadhwa suggests. For this reason, Catherine Rampell concludes that anxiety about automation always seems foolish in hindsight.[42]

How then to explain this anxiety? On this question, critics have been somewhat at a loss. For one, as Rampell notes, anxiety about technological unemployment is far less prevalent in other societies with mechanized economies, perhaps due to stronger labor protections. Furthermore, this anxiety has been present in times of economic boom and bust; it cannot simply be explained away as a response to economic recession. She muses that there may a Freudian element to this anxiety, insofar as it imagines that the technologies we have invented (or "fathered") are rendering us unnecessary or obsolete.[43] This narrative is clearly exemplified by an Associated Press (AP) story that begins, "For decades, science fiction warned of a future when we would be architects of our own obsolescence, replaced by our machines; an Associated Press analysis finds that the future has arrived."[44] In order to explain this anxiety, Parramore similarly offers the "Frankenstein complex"—a term coined by science-fiction author Isaac Asimov to describe the fear of being replaced by a robot.[45] This fear is exacerbated by the robot's docility, a point to which I will return shortly.

So what are critics missing? How might we make sense of this anxiety? What really makes this time different for those anxious about automation is not simply that there will soon be no human labor left to automate, but more precisely the character of the labor that is currently being automated: it is the labor of doctors, lawyers, and other highly skilled professionals. As Farhad Manjoo writes in *Slate*, "In the next decade, we'll see machines barge into areas of the economy that we'd never suspected possible—they'll be diagnosing your diseases, dispensing your medicine, handling your lawsuits, making fundamental scientific discoveries, and even writing stories just like this one."[46] In an article in the *New York Times*, Steve Lohr echoes this analysis, lamenting the loss of the last repository of jobs: "white-collar business process jobs."[47] This last repository—consisting of cognitive forms of labor—is thought to have been compromised by advances in machine learning, robotics, and related technological developments: big data, cloud computing, smarter machines, and so on. As the AP reports, "For the first time, we are seeing machines that can think—or something close to it."[48] Thinking, in this context, involves pattern recognition, complex communication, and other cognitive functions typically thought of as human.[49] Brynjolfsson and McAfee provide a number of examples of thinking machines, including Google's driverless car and IBM's Jeopardy-playing computer Watson, now also used to assist doctors in diagnostics and care at Memorial Sloan-Kettering Cancer Center.

In fact, despite the attention-grabbing, histrionic claim that there will soon be no human labor left to automate, Brynjolfsson and McAfee suggest that certain forms of low-skill, low-wage work that require manual dexterity—for example, restaurant bussers, gardeners, chefs—are relatively safe from automation for the time being; this is another inconsistency in the literature, suggesting yet again that interpretation (rather than dialogic engagement) is required to make sense of this anxiety.[50] Supporting this argument, the AP notes that many of the new jobs created after the Great Recession were low paying jobs. Ford too suggests that "average" workers will likely be safer from automation than knowledge workers, especially those who are highly paid and whose work is relatively routine and relies on pattern identification.[51] He refers to this pattern of displacement as "top heavy," and predicts that high school graduates may prefer to compete over trade jobs rather than going to college in order to compete for knowledge jobs that will soon be automated, a prospect that deeply concerns Ford.

In addition to low-skill, low-wage work, Brynjolfsson and McAfee argue that creative labor is relatively safe from automation, insofar as machines are not good at ideation, whether artistic, entrepreneurial, or otherwise, and thus might provide a safe haven for jobs in the future.[52] The *Economist* similarly suggests that workers might increasingly gravitate to fields that require emotional labor: art, therapy and counseling, yoga instruction, and so on.[53] In an article for the *Atlantic*, Derek Thompson agrees that humans will always be better at caring for humans.[54]

This is not to imply that there is complete consensus about which job sectors are most at risk for automation. For example, in their influential 2013 article, "The Future of Employment: How Susceptible Are Jobs to Computerization?" Carl Frey and Michael Osborne argue that it is precisely high-skill, high-wage labor that is relatively safe from automation, while jobs in low-skill, low-wage fields are most at risk.[55] That said, journalistic treatments of technological unemployment overwhelmingly emphasize the threat of automation to skilled labor. In an article in the *Guardian*, for example, Tom Meltzer references Frey and Osborne's argument, but is more transfixed by an AP study that found that "almost all the jobs that had disappeared in the past four years were not low-skilled, low-paid roles, but fairly well-paid positions in traditionally middle-class careers."[56] This trend is characterized as especially troublesome insofar as it encroaches upon sacred occupational terrain. As Meltzer writes, "Knowledge-based jobs were supposed to be safe career choices, the years of study it takes to become a lawyer, say, or an architect or accountant, in theory guaranteeing a lifetime of lucrative employment. That is no longer the case. Now even doctors face the looming threat of possible obsolescence."[57]

In his series for *Slate*, Farhad Manjoo similarly shifts from recognizing the current displacement of semi-skilled white-collar labor to ruminating on the possible future displacement of professional white-collar labor. Like Meltzer, Manjoo is unsettled: "Imagine you've spent three years in law school, two more years clerking, and the last decade trying to make partner—and now here comes a machine that can do much of your $400-per-hour job faster, and for a fraction of the cost. What do you do now?"[58] He later confesses, "As someone who likes his job (and his paycheck), what I saw terrified me."[59] This is not to say that the automation of low-skill work is completely ignored in journalistic treatments of technological unemployment. However, it is typically brought up as a way to introduce and frame the automation of professional labor. As Matt Miller asks in the *Washington Post*, "What if it's not just the 'unskilled' who are at risk, but most of us?"[60] Jaron Lanier similarly asks, "Is this the precedent that we want to follow for our doctors and lawyers and nurses and everybody else?"[61]

Why should the prospect of the disappearance of professional, white-collar jobs in particular be a cause for anxiety? Following Meltzer and Manjoo, one could argue that this attachment to skilled, professional labor is related to the paycheck it brings, or to the effort already expended in skill acquisition—the "hard-won expertise" of the professional class, as Meltzer characterizes it.[62] Indeed, it seems a waste of time and money to learn a trade that will soon be automated out of existence. More significantly, however, the attachment to skilled, professional labor appears to express a valuing of this labor as extraordinary in contrast to other kinds of labor seen as merely ordinary. As Meltzer writes, "For many, what were once extraordinary skillsets will soon be rendered ordinary by the advance of the machines. What will it mean to be a professional then?"[63]

What seems to distinguish extraordinary labor from ordinary labor here is not simply that extraordinary labor is scarcer than ordinary labor, but that it is "uniquely human"—a term used in the Pew survey referenced above—since it is the inimitability of this labor that renders it extraordinary. The uniquely human, in turn, is characterized by "higher" human faculties such as intellect, reason, and judgment—those qualities thought to be inimitable. It is according to this logic that scholars can imagine extraordinary labor as now threatened by "the advance of the machines." In contrast, ordinary labor is labor that can also be done by a machine (or an animal for that matter); the ordinary is thus characterized by "lower" human faculties associated with the body. In short, to be uniquely human is to have brains over brawn, and to be more than either machine or animal, where an association with nature or the body functions as a means of degradation, reproducing a Cartesian dualism between mind and body and their corresponding value attributions.

It is thus not that cognitive, white-collar labor more fully expresses the uniquely human, but rather that the relative historical insusceptibility of white-collar labor to automation provided a discursive basis for claiming the extraordinary, as if this insusceptibility were proof of the inherent value of cognitive labor. Unfortunately for those with a vested interest in the social status afforded to white-collar labor in part because of these associations, it is more or less an accident of history that manual labor was automated before cognitive labor, allowing manual labor's relative susceptibility to automation to serve as justification for the valuing of cognitive labor's "higher" faculties. As it turns out, cognitive labor is not more difficult to automate than manual labor; it simply required the advent of computers that could reproduce its particular patterns. As Lanier points out, computers can automate any patterned activity— manual or cognitive—regardless of the skill required.[64]

To reiterate, it is not that cognitive labor is valued because it is extraordinary or uniquely human, but rather that claiming it as extraordinary and uniquely human further legitimates a valuing of the mind and devaluing of the body. This is why the prospect of the automation of professional labor has been so unsettling, at least in part; what does one do when the discursive basis for legitimating one's values (and the relations of power these values support) is undermined? Not only is cognitive labor now subject to automation, but the very distinction between mind and body on which the differentiation between cognitive labor and manual labor depends is drawn into question, as by Hans Moravec's discovery that many forms of advanced reasoning require fewer computational resources than do basic sensorimotor skills.[65] If reasoning belongs to the mind, so too does control of the body, perhaps even more intensively insofar as bodily movement is the more resource-taxing skill.

The higher faculties of cognitive labor are valued not only as a means to elevate the mind over and above the body, but additionally as a function of their discursive proximity to what I will describe as willfulness; the "problem" with manual labor and its natural and artificial substitutes is that it is understood not only as more bodily, but in relation to this, as more docile. In contrast, cognitive labor is valued because it is understood not only as more intellectual but, in relation to this, more willful. Willfulness, in turn, is valued as a condition for political resistance and, alongside this, collective governance. In this way, I will argue, the anxiety surrounding the automation of cognitive labor can be understood as a response to a perceived threat to collective governance, as well as offering a solution to this threat by calling readers back to responsible forms of relationality.

* * *

To begin to elaborate these connections, it is instructive to review the history of the distinction made in political-economic thought between human labor and the labor of machines and of nature, insofar as the association of human labor with the exercise of collective governance is rooted in classical economics. Classical economists were first to claim a unique place for human labor in a theory of economic value. In particular, Adam Smith suggested that value stems from processes of production rather than exchange. His conceptualization of labor as the source of all value departed from that of physiocrats and utilitarians—pre-classical economists who theorized economic value as a product of relations of exchange; prior to these relations, they argued, there is only abundance or scarcity.[66] On this point Michel Foucault cites Étienne Bonnot de Condillac: "To say that a thing has value is to say that it is, or that we esteem it, good for some use. The value of things is thus founded on their utility, or, what amounts to the same thing, on the use we can make of them."[67] In their theorizations of value, physiocrats and utilitarians thus distinguished between value and wealth, proposing that nature produces wealth, which becomes valuable through exchange; outside of exchange, the wealth produced by nature has no value.

For classical economists this distinction proved somewhat problematic, owing to their proposition that labor, rather than exchange, produces economic value. How could classical economists explain why nature produces wealth but not value, and why only human labor produces value, when the sun and rain (for example) seem to labor as much as a farmer in producing a season's crops? How could they explain the economic value of an apple (for example), grown by the sun, rain, and earth, picked from a tree by a passerby, and sold in the market? The same point could be argued of the production of wealth by machines. For the classical economists, then, nature and technology posed a singular problem: the non-human production of wealth.

This problem must have been heightened by contemporaneous scientific understandings in biology, which drew the unity, coherence, and uniqueness of the human body into question. As Foucault notes in *The Order of Things*, the study of biological functions in the life sciences at the end of the eighteenth century precipitated a breakdown of the opposition between the living and the mechanical. He locates the origins of this breakdown (in part) in a shift in thought—from taxonomies of being, which focused on visible organs, to life sciences, which did not take for granted the visible similarities between organisms. If the identification of difference between organisms had previously relied on a visual study of their outward appearance, the dissection of organisms into their constitu-ent parts made possible an identification of their profound similarities.[68] This identification was furthered by the analytic organization of bodies and their parts in terms of functions, which revealed similarities masked by visual differences. Foucault argues that this commonality of bodily functions suggested the extent to which bodies are precisely mechanical. He concludes that by the end of the eighteenth century, the living could not be easily distinguished from the mechanical.

Similarly, as George Caffentzis has argued, the developing science of thermodynamics also suggested a symmetry between human and ma-chines vis-à-vis work.[69] This would present a problem for Marx (as for Smith): his iteration of the labor theory of value depended precisely on an asymmetry between humans and machines, such that only human labor could be thought to produce value. For this reason, Marx was anxious to distinguish between the labor of thermodynamics and the labor of political economy, as Caffentzis suggests, though in certain ways Marx's political-economic theories would mirror those of thermody-namics, as in his law of the conservation of value. However, Caffentzis contends, the proper context and center of Marx's theory of machines lies neither in thermodynamics nor biology, but rather in a political choice made in response to capitalist threats of automation, much like the threats levied against contemporary workers in the service indus-tries (that is, the threat that workers must be docile or else be replaced by machines). Caffentzis writes that Marx, like Thorstein Veblen, might easily have argued that machines can produce value insofar as they are a material expression of general social and scientific labor.[70] The reason Marx did not argue this is that he was invested in the willful organiza-tion and resistance of human labor: it is this property, I will explain, that truly distinguishes human labor from the labor of machines and nature in a Marxian paradigm.

Pondering the possibility that all human labor might one day be au-tomated, Caffentzis asks, "For if machines cannot create value, why then can labor?"[71] He reasons that the answer must not lie in any "positive" characteristic of labor: if a human can do it, a machine can also do it, at least in theory. That is, there is no activity that is inherently resistant to automation; it is only a matter of time until all activities can be automated. This leads Caffentzis to conclude that the value created by human labor must be a function of its "negative" qualities, or the way in which human laborers, unlike machines, can refuse to labor.[72] The classical economic distinction between labor and labor power is essential here: while nature and machines can perform work (they can labor), they cannot be said to have a capacity for work (they do not have labor power) insofar as they cannot refuse to work. As Jussi Vahamaki and Akseli Virtanen argue, potentiality is a uniquely human property, with its shadow of impotentiality, that is "the power not to pass into actuality."[73] This is rather more complicated than the definition of labor power Marx offers in *Capital*—"the aggregate of those mental and physical capabilities existing in a human being, which he exercises whenever he produces a use-value of any description"—and yet it captures more fully the essence of Marx's concept of labor power.[74]

As Caffentzis suggests, Marx's affirmation of labor power and its value as distinctly human was motivated by the possibility of refusal, and more specifically of labor's refusal of the extraction of surplus value by capitalists and of the unjust organization of labor under capitalism. The exceptional status of human labor has thus long been tied to notions of willful refusal and resistance, with technology serving as its docile foil. This is not to suggest that labor is not in fact willful, but rather that its capacity for willfulness becomes its defining property.[75] In fact, it is because of labor's capacity for willfulness that capitalists threaten workers with automation as a strategy to manage and discipline them. In order to make automation less attractive and thereby forestall their own displacement, workers are told that they have to be less costly, less demanding, and more productive.

This disciplining is exemplified by a television advertisement released in early 2014 by the Employment Policies Institute, a lobbying front for the hotel, restaurant, alcohol, and tobacco industries.[76] The ad depicts a series of mundane retail scenes: a gas station attendant pumps gas; a cashier scans grocery items; a waiter takes customers' orders in a restaurant. In each of these scenes, the worker slowly fades from view, and the driver is left to pump her own gas, the shopper to scan her own grocery items, and the restaurant customers to use a touchscreen device to place their orders. A voiceover narrates, "President Obama wants to raise the minimum wage by nearly 40 percent. That may sound like a good idea, but if customers won't pay for it, it forces employers to install technology that takes the place of entry-level jobs. Every time you use a self-checkout lane or even a touch screen ordering system, it's a task that used to be part of someone's job description. When you raise the minimum wage, a new government report confirms that up to one million jobs will disappear."[77] If this narrative is most often deployed by capitalists, or rather their PR firms, it has also taken root in popular consciousness, as expressed in journalistic coverage of automation. To take one example, in one of the first post-recession articles on technological unemployment, Manjoo begins with a provocation:

> If you're taking a break from work to read this article, I've got one question for you: Are you crazy? I know you think no one will notice, and I know that everyone else does it. Perhaps your boss even approves of your Web surfing; maybe she's one of those new-age managers who believes the studies showing that short breaks improve workers' focus. But those studies shouldn't make you feel good about yourself. The fact that you need regular breaks only highlights how flawed you are as a worker. I don't mean to offend. It's just that I've seen your competition. Let me tell you: You are in peril.
>
> At this moment, there's someone training for your job. He may not be as smart as you are—in fact, he could be quite stupid—but what he lacks in intelligence he makes up for in drive, reliability, consistency, and price. He's willing to work for longer hours, and he's capable of doing better work, at a much lower wage. He doesn't ask for health or retirement benefits, he doesn't take sick days, and he doesn't goof off when he's on the clock.
>
> What's more, he keeps getting better at his job. Right now, he might only do a fraction of what you can, but he's an indefatigable learner—next year he'll acquire a few more skills, and the year after that he'll pick up even more. Before you know it, he'll be just as good a worker as you are. And soon after that, he'll surpass you.
>
> By now it should be clear that I'm not talking about any ordinary worker. I'm referring to a nonhuman employee—a robot, or some kind of faceless software running on a server.[78]

Here we might recall that the word "robot" has roots in the Czech words "robota" (drudgery, servitude, forced labor) and "robotnik" (slave). Like a slave, the worker's robot replacement is figured as utterly controllable, unlike the demanding, lazy worker Manjoo conjures in the passage above.

It is telling that so much of the discourse surrounding technological unemployment centers on the notion that robots "steal" jobs, or if not robots, then some other discrete technology adopted in lieu of a human laborer. However, as Brynjolfsson and McAfee argue, technological unemployment typically results from the restructuring of work processes in a way that eliminates jobs, rather than a robot literally being employed in a human's stead. For example, as the *Economist* notes, the Industrial Revolution did not simply entail the displacement of human labor by machine labor, but rather a comprehensive reshaping of jobs in concert with technological advances.[79] The mischaracterization of technological unemployment as a one-for-one displacement of workers by machines highlights the extent to which the robot is important precisely as a symbolic figure: the docile foil to a willful worker.

For conservatives and progressives alike, work is a particularly receptive institution for establishing a discursive boundary between docile machines and willful humans insofar as work is structurally imposed and thus requires willfulness to resist. Indeed, willfulness is often associated with refusal and resistance, or what Sara Ahmed calls "the not": "The will can be rearticulated in terms of the not: whether understood as possibility or capacity, as the possibility of not being compelled by an external force . . . or the capacity to enact a 'no' to what has been given as instruction. Indeed, willfulness as a judgment tends to fall on those who are *not* compelled by the reasoning of others. Willfulness might be what we do when we are judged as being *not*, as not meeting the criteria of being human, for instance."[80] The association of willfulness with resistance (including, as Ahmed suggests elsewhere, unwilling obedience) has made willfulness appealing to intellectual and political traditions on the Left that value resistance. This includes Marxist traditions that value the resistance of exploited laborers and feminist traditions that value the resistance of willful women—figures whose refusal to go along with the general will marks their particular wills as pathological. As Ahmed notes, the attribution of willfulness is often meant to identify a person with a problem; willfulness is a fault of character.[81] Those with power, on the other hand, are rarely seen as willful insofar as their will has been institutionalized.[82] We have already seen how capital leverages the charge of willfulness at labor, characterizing demands for higher wages as being against the general will, which is expressed symbolically through the figure of the economy. As Ahmed notes, workers who strike become a blockage of sorts, impeding the financial flows of the economy.[83]

If willfulness has been linked to resistance, a kind of detachment from the general will, it has also been historically linked to accountability, a kind of attachment. Ahmed traces this treatment of willfulness to Nietzsche, for whom the attribution of willfulness is central to the process of becoming an accountable subject; this attribution works to unify a subject who can be held accountable for his or her actions, who can be made guilty.[84] Will, or more precisely willpower, becomes a requirement for the development of a responsible, moral subject. Following Nietzsche's conceptualization of will (if only for a moment), Ahmed thus characterizes attributions of willfulness as a "straightening device." But rather than leading her to reject attributions of willfulness as disciplinary, this characterization provides her with a foundation for reexamining and investing selectively in the political possibilities of such attributions—in other words, for queering willfulness. She writes, "If we are charged with willfulness, we can accept and mobilize this charge. To accept a charge is not simply to agree with it. Acceptance can mean being *willing to receive*."[85] In this way, Ahmed comes to value willfulness (if ambivalently at times) as a possible locus of radical disobedience.

In queering the will (or theorizing the will's queer potential), Ahmed departs from Nietzsche's account as well as from feminist and poststructuralist critiques of the will as masculinist or just generally outmoded. Instead, she takes a note from ancient Roman philosopher Lucretius in arguing that neither the will nor willfulness need belong to the subject. For Ahmed, the will often escapes efforts to contain it within a subject; this is its perverse potential.[86] Indeed, Ahmed locates willfulness in many other places: in body parts (arms play a particularly central role in the book) as well as non-bodily matter (stones, for example). Rather than positing the will as an occasion for subjection, as Nietzsche does, Ahmed thus provocatively reimagines willful resistance without

subjection or intentionality. In her account, willfulness doesn't necessarily belong to the subject; rather, it belongs to particular experiences
of willfulness that can exceed subjection.[87]

That said, Ahmed seems ambivalent about leaving the (social) subject
behind; the book is called *Willful Subjects* after all. In a passage that exemplifies this ambivalence, Ahmed offers an interpretation of Antonio
Gramsci's oft-quoted phrase "optimism of the will, pessimism of the intellect." She notes that for Gramsci, the efficacy of willful resistance requires effort and work. As she quotes Gramsci: "It should be noted that
very often optimism is nothing more than a defense of one's laziness,
one's irresponsibility, the will to do nothing. It is also a form of fatalism
and mechanicism."[88] Ahmed interprets Gramsci here as arguing that our
willfulness must be active, intelligent, oriented toward the future, and—
crucially—collective. For willfulness to be effective, Ahmed suggests
(following Gramsci), we must work with others toward the realization
of collective visions and goals.[89]

This call to the social is particularly pronounced in the final chapter
of the book ("Willfulness as a Style of Politics"), in which Ahmed cel-
ebrates Rosa Parks's willfully resistant contribution to the Civil Rights
movement, in addition to considering Reclaim the Night marches,
the Occupy movements, and a protest against a local news station for
refusing to air Marlon Riggs's 1989 film *Tongues United* (described by
Eve Kosofsky Sedgwick in *Touching Feeling*). In reference to Occupy,
Ahmed references Bartleby the Scrivener's famous refusal, "I prefer not
to"—noting that the word "prefer" is described in the novel as "queer."[90]
But far from simply preferring not to, Ahmed's "feminist killjoys" (a
reclaimed insult) are called upon to "intrude on a world in which we
figure as intrusion," "to assert our existence in order to exist," and to
"find voices."[91] Willfulness here seems to take the familiar, one might say
normative, form of collective, democratically oriented, political action
rather than a queer refusal to participate or, if refusal is always already
embedded in the democratic project, a kind of apathetic or lazy disinclination to participate—what Leo Bersani refers to as "inaptitude."

Might Ahmed's analysis accommodate an antisocial willful subject or
non-subject? Despite Ahmed's valuing of collective resistance, she notes
that willfulness "can be a way of withdrawing from the pressures of an
oppressive world and can even become part of a world-making project.

Willfulness as a diagnosis can thus be willingly inhabited, as a way of creating a room of one's own."[92] She elaborates:

> Willfulness is ordinary stuff. It can be a daily grind. This is also how an experience of willfulness is world creating: willful subjects can recognize each other, can find each other, and can create open spaces of relief, spaces that might be breathing spaces, spaces in which we can be inventive. If in most spaces we have to be assertive just to be, we can create spaces which give us freedom from that necessity. There can be joy in creating worlds out of the broken pieces of our dwelling spaces: we can not only share our willfulness stories, but pick up some of the pieces too. And we can hear each other in each other: can be moved by each other with each other, we can even just tell each other to let it go, at the moments when holding on demands too much. We can say this, as we have been there, in that place, that shadowy place, willful subjects tend to find themselves; a place that can feel lonely can be how we reach others.[93]

While Ahmed puts these two ends side-by-side—a room of one's own and the project of world-making—we might also contrast them, insofar as a room of one's own might be asocial or antisocial, a way of leaving the collective without making it anew. As Ahmed argues elsewhere, "A queer relation offers the freedom of not having a relation, the freedom not to participate, not to be connected or stay connected."[94] In *Willful Subjects* and other texts, Ahmed makes a case for straying from that which would straighten the queer, but this turning away from one iteration of the social often seems to entail creating it anew elsewhere. Again, we can ask here if the will to disconnect that Ahmed describes might also include the refusal of all forms of social subjection, including that of the feminist killjoy. Gramsci admonishes laziness, irresponsibility, and the "will to do nothing," but are not these too forms of resistance, that is of resisting the social?

The ambivalence of *Willful Subjects* with regard to antisocial forms of willfulness is instructive, illuminating the ways that the association of the human with willful refusal or resistance is not simply about saying "no" or being a "not," but is often about saying "yes" to other forms of subjection, if not without some hesitation or qualification. My aim here is not to oppose a queering of the will—quite the opposite: to truly queer the will one would need to recognize fully that which has been associated with willfulness, its discursive baggage. If, for the Left, the willfulness of labor is expressed through collective resistance, this willfulness embraces a different kind of social subject: a laboring subject who says no to exploitation and alienation but must also say yes to collective governance.[95] The subject is thus not abandoned but reformed.

This is not to say that the Left—with its investment in collectively self-governing laboring subjects—has been universally anxious about automation or technological unemployment. As noted above, some on the Left (particularly the far Left) reject the notion that automation constitutes a threat to labor, and instead conceptualize technological development as a capitalist strategy to gain temporary advantages over competitors in the market; it is the capitalists, not the "robots," that are the problem. Rather, my aim here is simply to point out that contemporary anxieties about automation and technological unemployment are grounded in an attachment to the human as a social subject—and more specifically, a collectively self-governing laboring subject—which finds expression through an association of the human with willfulness, a discursive inheritance from classical political economy.

Insofar as cognitive/professional labor is narrated as uniquely human, it is discursively nearest to this attachment. It is for this reason, I am proposing, that the prospect of the elimination of this kind of labor is particularly distressing.[96] Unlike capitalists' ascription of willfulness to labor, which is mobilized to manage labor's demands and make labor more docile—more like the machines that will "steal" our jobs if we do not become more docile—the anxious investment in cognitive/professional labor as willful is mobilized toward the production of a less docile, more responsible subject. In other words, the anxiety that surrounds the automation of white-collar labor does not aim to make labor more docile, but rather to make us all more willful, where willfulness provides a foundation for collective governance.

This does not mean that white-collar labor is in reality less docile, more resistant, more difficult to manage, or more collectively oriented than blue-collar labor—labor history suggests quite the opposite—but rather that it is symbolically so. This symbolism informs the fantasy of collective governance, with its investment in a society constituted by mindful, rational, and, perhaps above all, responsible subjects. Of what use could blue-collar labor be to such a project, with its own particular web of meanings—the animal, the bodily, the emotional, and the libidinous?[97] If the automation of blue-collar labor has been a more acute source of anxiety in the past, this anxiety must be read differently than the anxiety that currently surrounds the automation of professional, cognitive labor. For example, it might be read as an anxiety about flagging masculinity (insofar as blue-collar labor is circumscribed by a series of gendered meanings), or about the indolence or irresponsibility of unemployed laborers (insofar as blue-collar labor is also circumscribed by a series of raced and classed meanings), and so on. In expressing anxiety about the automation of cognitive labor, the texts examined here mean differently, and through these meanings they work to hail a reader into the responsible subject position "threatened" by automation.

While I have focused in this chapter on the anxiety surrounding automation and technological unemployment, it is instructive to consider briefly one imagined transformation that engenders hope in scholars and journalists writing about technological unemployment, insofar as it adheres to the value ascription here problematized. The attachment to a collectively self-governing laboring subject is expressed not only in the fear that the advance of technology will render cognitive labor obsolete, but also in the notion that automation might transform work to be less repetitive and routine and thus more interesting, satisfying, and fulfilling. As the Pew survey cited above notes, some tech experts are rather hopeful about the state of future employment, because they believe that automation will make possible the invention of new types of work that take better advantage of quintessentially human capabilities and that will therefore be less tedious and more enjoyable and socially beneficial.[98] As Brynjolfsson and McAfee suggest, when machines perform work that is easily routinized, humans are freed up to do less repetitive work and more creative, interactive work.[99] Here, automation is imagined as a benevolent force, eliminating tedium and inviting workers to identify positively with their work.

As with the notion that human labor is uniquely willful, the idea that the institution of work might be transformed so substantially as to render it liberatory (rather than oppressive) has roots in classical economics. As Brynjolfsson and McAfee cite Adam Smith: "The man whose whole life is spent performing a few simple operations, of which the effects are perhaps always the same, or very nearly the same, has no occasion to assert his understanding."[100] Again, it is the automation of routine forms of labor that frees up laborers to engage in work that would allow them to assert their understanding and, thereby, to identify with and express themselves through their labor. On this point we might also consider a famous passage from Marx and Engels's *German Ideology*:

> The division of labour offers us the first example of how, as long as man remains in natural society, that is, as long as a cleavage exists between the particular and the common interest, as long, therefore, as activity is not voluntarily, but naturally, divided, man's own deed becomes an alien power opposed to him, which enslaves him instead of being controlled by him. For as soon as the distribution of labour comes into being, each man has a particular, exclusive sphere of activity, which is forced upon him and from which he cannot escape. He is a hunter, a fisherman, a herdsman, or a critical critic, and must remain so if he does not want to lose his means of livelihood; while in communist society, where nobody has one exclusive sphere of activity but each can become accomplished in any branch he wishes, society regulates the general production and thus makes it possible for me to do one thing today and another tomorrow, to hunt in the morning, fish in the afternoon, rear cattle in the evening, criticise after dinner, just as I have a mind, without ever becoming hunter, fisherman, herdsman or critic. This fixation of social activity, this consolidation of what we ourselves produce into an objective power above us, growing out of our control, thwarting our expectations, bringing to naught our calculations, is one of the chief factors in historical development up till now.[101]

In this passage, Marx and Engels call for the reorganization of labor from the "natural" (that is, forced) to the "voluntary." In this transformation, that which "enslaves" laborers is not left behind or abandoned, but reclaimed through an assertion of control, presumably the control of collective governance. As Marx and Engels write, "Each can become accomplished in any branch he wishes," but the branches remain forms of work: hunting, fishing, herding, and—thank goodness—cultural criticism. In this vision, the institution of work remains central to the project of being human, not simply because work is necessary to survive, but because it is necessary for the survival of the social. To be human is to work, and to work is to be of use to society. As laborers we have purpose—not simply something to do, but something to contribute. Conversely, without work, we become less than human.

The notion that automation and technological unemployment portend the obsolescence of the human reveals the extent to which labor, particularly cognitive labor, is still seen as central to the project of being human. As in Marx and Engels's vision, the contemporary call to identify with and exert control over one's labor is coupled with a promise that work might be less tedious and more satisfying, interesting, and fulfilling, not simply because workers will be allowed to indulge diverse desires, but because these desires will remain oriented toward the social. For this to happen, though, workers need to exert control, which is also to say they need to become responsible, whether through the state or more directly through collective governance.

If it is true, as I am arguing here, that it is not (or not simply) real white-collar laborers and their real incomes that are at stake for those anxious about automation and technological unemployment, but rather the symbolic white-collar laborer, whose value lies in "his" status as an intellectual, willful, and collectively oriented self-governing subject, then it stands to reason that the anxiety surrounding automation and technological unemployment is not primarily about how to fix the income issue, but rather is itself a kind of fix to a different issue: the waning of valued relational bonds. That is to say, this anxiety is less about the workers whose jobs/incomes are supposedly imperiled by automation than it is about these texts' readers, who are solicited to form or maintain a symbolic attachment to professional labor, whether as that which needs to be protected by the state (acting on behalf of the social) or as that which needs to self-organize willfully against the forces of automation and/or capital. In both cases, this solicitation constitutes a call to governance, which becomes necessary to protect the social and is itself an agent of the social.

## An End to Work

An article in the satirical news outlet the *Onion* titled "Chinese Factory Workers Fear They May Never Be Replaced with Machines" begins,

"Expressing growing concerns about their future job security, factory workers across China reported this week that they are deeply worried that they may never lose their menial, hazardous positions on product assembly lines to automated machinery."[102] Trust the satirist to seize upon a truth that so many scholars and journalists concerned about technological unemployment seem determined to ignore. One need not be an exploited factory worker laboring in hazardous conditions to appreciate the sentiment that employment is hardly a panacea, and that losing one's job as a result of automation might at the very least be a double-edged sword.

Apart from work's attachment to the wage, why does the prospect of an end to work engender anxiety, and what does this anxiety mobilize or make possible? This chapter has argued that at the broadest level, contemporary anxiety surrounding technological unemployment serves as a call to action as well as a call that is itself a kind of action. The final line of Brynjolfsson and McAfee's book is instructive here. They write, "Technology is not destiny. We shape our destiny."[103] Similarly, the Pew survey referenced above identifies as a key theme (among the experts questioned about automation and technological unemployment) that "ultimately, we as a society control our own destiny through the choices we make."[104] More specifically, this call to action (and action through call) is aimed at preserving the social as established through the institution of work, whether through state or collective governance. Our power to "shape" and "control" our destiny "through the choices we make" thus appears to be heavily circumscribed by a valuing of collective relationality. The human agent that takes up this project of shaping and controlling is also quite specific. To be human is not simply to have a will or to be willful, but to express this will through work, particularly professional/cognitive work, the value of which is asserted through the anxiety surrounding technological unemployment.

This argument helps to explain why Lanier, for example, fuses together the social contract established through relations of work with his conception of the human; linking these is a way of valuing the social insofar as the "human" (like the "natural") is discursive shorthand for both the inevitable and the important. As Lanier writes at the beginning of *Who Owns the Future?*, "This is a book about futuristic economics, but it's really about how we can remain human beings as our machines become so sophisticated that we can perceive them as autonomous."[105] For Lanier, remaining human seems to require that we participate through work in a social contract that preserves the middle class and, thereby, democracy. In short, to be human is to be a collectively oriented social subject.

This valuing of work as a social institution, rather than simply an economic institution, is also made apparent at the end of *The Second Machine Age*, when Brynjolfsson and McAfee cite Voltaire: "Work saves a man from three great evils: boredom, vice, and need."[106] Taking inspiration from Voltaire, they write: "It's tremendously important for people to work not just because that's how they get their money, but also because it's one of the principal ways they get many other important things: self-worth, community, engagement, healthy values, structure, and dignity, to name just a few."[107] They continue, "Whether the focus is on the individual or the community, the conclusion is the same: work is beneficial."[108] For this reason, Brynjolfsson and McAfee reject the guaranteed basic income as a method of ensuring mass consumption in times of high unemployment or underemployment; according to them, it would not sufficiently encourage or reward work, which is important not simply for a "bounteous economy" but a "healthy society."[109] While Ford advocates for income to be issued by the state as a reward for pro-social behavior, he is similarly skeptical of the guaranteed universal income, which offers "no motivation for self improvement, no sense of self-worth and no hope for a better future."[110] He calls avoiding work in favor of government support a "moral hazard" and argues that people need incentives that ensure that they "do what is best for themselves and for society as a whole."[111] As he writes, "If we cannot pay people to work, then we must pay them to do something else that has value."[112]

According to these arguments, people must work not simply to eat, but to be happy and, perhaps most importantly, to be good.[113] To this end, Brynjolfsson and McAfee cite a number of studies that show a correlation between employment and societal health, as measured through marriage/divorce and crime rates, inadvertently identifying work as an institution with strong ties to social control. Similarly, Ford writes that jobs "provide a useful occupation for our time" as well as "hope for advancement" necessary for both individual and social stability; motivate us to invest in various forms of self-improvement; and give us a "sense of purpose" and "a more orderly and civil society."[114] The more these authors imagine human labor might truly be unnecessary or simply not viable in the future, the more their arguments shift from the economic to the social and ethical/moral. In their conclusion, for example, Brynjolfsson and McAfee speculate that "as more and more work is done by machines, people can spend time on other activities. Not just leisure and amusements, but also the deeper satisfactions that come from invention and exploration, from creativity and building, and from love, friendship, and community."[115] Here leisure and amusement are quickly dismissed as less satisfying than forms of activity that require work and maintain social bonds. Values play a central role in this maintenance. As they write at the end of their conclusion, "Our success will depend not just on our technological choices, or even on the coinvention of new organizations and institutions. As we have fewer constraints on what we can do, it is inevitable that our *values* will matter more than ever."[116] They continue, "In the second machine age, we need to think much more deeply about what it is we really want and what we value, both as individuals and as a society."[117]

In turning to values, Ford, Brynjolfsson, and McAfee take a note (if inadvertently) from John Maynard Keynes, who coined the term "technological unemployment" in his 1930 essay "Economic Possibilities for our Grandchildren." In this essay, Keynes asks, "What can we reasonably expect the level of our economic life to be a hundred years hence?"[118] For the sake of argument, he imagines humanity as eight times better off in this future. In fact, he argues, the "economic problem"—that is, the "struggle for subsistence"—will be solved, barring any wars and population booms.[119] However, the elimination of the economic problem will create a new problem, Keynes writes, which is that "mankind will be deprived of its traditional purpose."[120] He elaborates, "Thus for the first time since his creation man will be faced with his real, his permanent problem—how to use his freedom from pressing economic cares, how to occupy the leisure, which science and compound interest will have won for him, to live wisely and agreeably and well."[121] Pondering the prospect of this freedom, Keynes anticipates a "nervous breakdown" of all those who "have been trained too long to strive and not to enjoy."[122] He imagines that people will continue to continue to work for a while in order to be content, "but beyond this," he writes, "we shall endeavour to spread the bread thin on the butter—to make what work there is still to be done to be as widely shared as possible. Three-hour shifts or a fifteen-hour week may put off the problem for a great while. For three hours a day is quite enough to satisfy the old Adam in most of us!"[123]

In addition to this transformation of our activity, and perhaps more significantly, Keynes imagines a transformation of morality:

> I see us free, therefore, to return to some of the most sure and certain principles of religion and traditional virtue—that avarice is a vice, that the exaction of usury is a misdemeanour, and the love of money is detestable, that those walk most truly in the paths of virtue and sane wisdom who take least thought for the morrow. We shall once more value ends above means and prefer the good to the useful. We shall honour those who can teach us how to pluck the hour and the day virtuously and well, the delightful people who are capable of taking direct enjoyment in things, the lilies of the field who toil not, neither do they spin.[124]

Tellingly, Keynes embraces an end to work, but precisely because he imagines that the absence of work will bring about a return to values rather than threatening them, as Brynjolfsson, McAfee, and Ford fear. The only dread Keynes seems to feel is for the "ordinary man" whose habits and instincts related to work "have been bred into him for countless generations" and "which he may be asked to discard within a few decades."[125] Despite this inversion, it is significant that Keynes's diagnosis of technological unemployment occasions a reinscription of values, revealing once again that work matters discursively as a symbolic object through which sociality can be established; on this point he is perfectly aligned with Brynjolfsson, McAfee, and Ford.[126] In other words, it matters less whether actual work is valued, insofar as work is a conduit or proxy for valued forms of relationality. Whether by work or by "the art of life" a similar end is sought: virtue and goodness, particularly in relation to self-indulgent greed and other-directed altruism.

What might we stand to gain from the reduction or even elimination of work? Frances Fox Piven and Richard Cloward argue that the institution of work has deep ties to social control. They write, "The regulation of civil behavior in all societies is intimately dependent on stable occupational arrangements. So long as people are fixed in their work roles, their activities and outlooks are also fixed; they do what they must and think what they must. Each behavior and attitude is shaped by the reward of a good harvest or the penalty of a bad one, by the factory paycheck or the danger of losing it. But mass unemployment breaks that bond, loosening people from the main institution by which they are regulated and controlled."[127] In fact, the things that Piven and Cloward want to detach from work—people's "activities and outlooks"—are the very same things that the writers and thinkers examined here want to keep attached to work, or to some institutional replacement in the event of total automation. To be free from work is thus to be free from forms of regulation and control that are attached to work, and more broadly to loosen the social ties that make possible regulation and control. Not coincidentally, it is also to give oneself over leisure, or whatever non-work activity might become in the absence of work as a social institution. As Arthur C. Clarke is reported to have said, "The goal of the future is full unemployment so we can play"—not invent, build, love, commune, or the other high-minded values Brynjolfsson and McAfee list, but play.[128] Clarke concludes, "That's why we have to destroy the present politico-economic system."[129]

# 4

## Sharing

Where someone manages to commercialize a tribe's gift
relationships, the social fabric of the group is invariably
destroyed. . . .
If you want out, you pay your own way.
—Lewis Hyde, *The Gift*

In a promotional clip for the odd-job service TaskRabbit, Jennifer G. (identified only by first name and last initial, as are all "taskers" on the site) excitedly describes the first task for which she was hired—purchasing plane tickets within a budget for a trip through several different U.S. cities.[1] "They had a deadline," she says, "and as soon as the deadline hit I got an email that I got the. . . ." She pauses, seeming to catch herself before saying "job," and revises: "that he accepted my offer." This happens quickly in the clip and is far from the point of Jennifer G.'s story, but this difference—between a task and a job—is a significant feature of the "sharing economy," a term that gained traction in the early 2010s to describe markets for goods and services provided by amateurs rather than professionals or formal businesses, and typified by companies like TaskRabbit, Airbnb (for lodging, sometimes cited as the first sharing service), and Uber (for transportation). Rather than going to a hotel, the sharing economy invites users to stay with a "local" renting out a room in his or her apartment, to catch a ride with somebody with a car and time to spare, to borrow tools from a neighbor, and so on—typically for a price cheaper than (or at least competitive with) what one would pay otherwise.

The sharing economy offers a number of distinct advantages according to its advocates: consumers buy less, waste less, and are less encumbered by their possessions; unemployed, underemployed or otherwise financially struggling workers can boost their income and sense of self-worth; and community bonds, or at least casual social bonds, are strengthened. For example, Rachel Botsman and Roo Rogers write in their celebratory 2010 tome *What's Mine Is Yours: The Rise of Collaborative Consumption* (their term for the sharing economy):

> Collaborative Consumption is enabling people to realize the enormous benefits of access to products and services over ownership, and at the same time save money, space, and time; make new friends; and become active citizens once again. Social networks, smart grids, and real-time technologies are also making it possible to leapfrog over outdated modes of hyper-consumption and create innovative systems based on shared usage such as bike or car sharing. These systems provide significant environmental benefits by increasing use efficiency, reducing waste, encouraging the development of better products, and mopping up the surplus created by over-production and -consumption. . . . Collaborative Consumption is not a niche trend, and it's not a reactionary blip to the 2008 global financial crisis. It's a growing movement with millions of people participating from all corners of the world.[2]

Like Botsman and Rogers, Lisa Gansky advocates enthusiastically for the sharing economy, which, in a bid to put her own stamp on the phenomenon, she terms "the mesh." She writes, "The Mesh is that next big opportunity—for creating new businesses and renewing old ones, for our communities, and for the planet. And it's just beginning."[3] This quasi-utopian rhetoric is echoed in the promotional materials of Peers, a trade organization whose members include TaskRabbit, Lyft, and Airbnb (which paid for the organization's founding), as well as dozens of smaller companies.[4] A video produced by the organization begins, "Something revolutionary is happening," while a camera follows around a multiracial, multiethnic group of interconnected people with smiles plastered on their faces: a hiply dressed woman drops off a child at daycare before giving a walking tour, and then tends to what looks to be a community garden;[5] she waves to a man who is also gardening and who then borrows a car to deliver fresh vegetables from the garden to another man, who prepares a meal for a group of people that includes the gardener, and so on. These idyllic scenes of community life are set to the kind of inoffensive, upbeat instrumental music that has come to characterize advertisements for smartphone applications, while a narrator extols the virtues of the sharing economy, concluding, "Together, as peers, we are building the sharing economy, an economy with humanity at its center and community at its core."

A backlash to the sharing economy gained momentum in 2013, following class-action suits filed against Uber in August and rival Lyft in September. While the sharing economy has rarely if ever been cited as an example of technology "stealing" jobs—if anything, the opposite is the case—criticism of the sharing economy, like scholarship anxious about automation and technological unemployment, has similarly characterized the sharing economy as bad for workers and good for the Silicon Valley companies that stand to benefit financially from workers' impoverishment. Critics see the technologies used to coordinate peer-to-peer markets not as taking jobs, but rather as helping to create bad jobs that unemployed and underemployed laborers have little choice but to take. As Trebor Scholz writes, "The Internet has become a highly efficient enabler of unethical work arrangements."[6] Technology is less the culprit here than Silicon Valley companies, or rather those few employees who get rich while sharing-economy laborers are essentially forced to drive strangers around and rent out their apartments to make ends meet. While technology may not be seen as the culprit, it is nevertheless understood as the foundation of the sharing economy, insofar as coordinating the markets in question presents a logistical problem (that is, how to facilitate exchange between strangers), which is solved through digital network technologies. These technologies, in turn, are cast as yet another weapon of exploitation in the capitalist arsenal. For example, Rob Horning writes, "The sharing economy epitomizes the deployment of technology to intensify inequality, in this case by creating monopolies that aggregate and co-opt the effort and resources of many users, who are pitted against one another within the platforms."[7] Or as Stanley Aronowitz puts it, "These are not jobs, jobs that have any future, jobs that have the possibility of upgrading; this is contingent, arbitrary work. . . . It might as well be called wage slavery in which all the cards are held, mediated by technology, by the employer, whether it is the intermediary company or the customer."[8]

The sharing economy is thus thought to contribute to the increasing "precariousness" of labor, a term widely associated with autonomist Marxist thought (a strain of Marxism developed in Italy in the 1960s and

1970s), which describes the fragmentation and discontinuity of work relations and experiences in contemporary capitalism.[9] While precarious labor is a mixed blessing for autonomists (more on this later), scholars and journalists writing about precarious labor within the context of the sharing economy are almost exclusively critical. Their two primary objections are that (1) sharing-economy labor represents a regression from steady, full-time employment, and its wages, benefits, and job security; (2) sharing-economy labor is structurally coercive in that workers have essentially been forced into laboring for the sharing economy as a direct result of post-recession unemployment and underemployment.

To begin with the first objection, critics contend that sharing-economy labor practices represent a regression of labor standards across a number of fronts: wages are lower, benefits like health insurance and retirement plans are virtually nonexistent, taxes and necessary expenses are paid out-of-pocket, promotion is rarely an option, mentoring is scarce, and perhaps most importantly, workers have little if any job security. This critique has been informed and bolstered by ethnographic data, gathered primarily through interviews with laborers, though also in a few cases through participant observation. One of the most thorough descriptions of sharing-economy labor appears in Sarah Kessler's autoethnographic, investigative report for *Fast Company*, "Pixel and Dimed: On (Not) Getting By in the Gig Economy." In this article, Kessler describes a month of looking for and sometimes finding work in the sharing economy, with the goal of surpassing a raised minimum wage ($10.10 an hour) and on the condition that she must accept any work she is offered.[10] Kessler searches far and wide to find work: offering herself as a proofreader on Fiverr (one of about 4,786 similar ads, no takers), a pizza chef on Kitchensurfing (no takers), a gift-wrapping instructor on Skillshare (no takers), and a dog walker on DogVacay. She also submits applications to Zirtual (not hiring), FancyHands (not hiring), ChaCha (not hiring), Exec (rejected), WunWun, and Postmates.

At an orientation for Postmates—an on-demand delivery service—Kessler learns that the two thousand delivery people who work for Postmates are not considered employees and do not receive benefits, unlike the forty-five engineers, designers, and executives who are considered employees by the company. She also notes that delivering for Postmates is insecure; workers must compete with each other for delivery jobs, and when there is not enough work to go around, it is the workers rather than the company that suffers. Similarly, when workers get flat tires or face other maintenance issues, it is their problem, rather than the company's. For these reasons, it seems impossible for workers to make enough money to survive by working for Postmates alone; Kessler talks to one delivery person who says that Postmates is not a main source of income for him or for anyone he knows. Because Kessler does not own a car, she is discouraged from pursuing the job.

Kessler has the most luck with TaskRabbit, though finding work through TaskRabbit also proves difficult. Even very low-paying tasks are quickly allotted to other taskers before Kessler has a chance to apply. She makes $8.80 an hour helping somebody open mail, before deciding to hire a fellow tasker to mentor her in landing tasks. The person she hires—Dmitry Solominsky—is described in Kessler's article as a rare success story, and serves as a foil to the average sharing-economy laborer:

> For every Solominsky I meet, I can easily find dozens of people like Sharon in San Diego, who has a goal of making $300 a week on TaskRabbit to help pay her bills, but hasn't hit it yet. Or Kristen in New York City, who bids on tasks when she's working full-time as a receptionist. Or Stacie, who works full-time as a software engineer in Boston, but always keeps the TaskRabbit website open so she can complete tasks on her lunch hour, after work, on weekends, or without leaving her desk. Stacie made about $6,000 on TaskRabbit last year, earning her "elite TaskRabbit" status. She likes helping people out, but she would never work on TaskRabbit just for the money. "If I wasn't working full time, I could do more tasks," she tells me, "but even if I doubled that, that's still poverty—$12,000 a year. And there are no benefits. You don't know what you're going to wake up to. You could wake up one day, and be like, oh my god, I made $300 today, and then have three days where you're making $12."[11]

After her mentoring session with Solominsky, Kessler is hired (for $20 an hour) by a frustrated mother who needs assistance in making her thirteen-year-old daughter do her homework (Kessler writes that nothing makes her want her desk job back more than this task), to participate in a flash mob (for $20 total including a two hour rehearsal), to wrap gifts (for $20 an hour), and to audition as a personal assistant (for $15 an hour). Eventually desperate to earn more money, Kessler looks for work through Amazon's low-paying service Mechanical Turk, taking a twenty-four minute survey (for 70 cents) and labeling photographs (for $1.94 an hour).

Apart from simply finding work, which takes up a fair amount of time, one of Kessler's primary difficulties is scheduling so many small tasks, which often conflict with each other, as well as commuting between tasks. While the sharing economy is sometimes touted as accommodating tricky schedules, Kessler feels that she actually has little control over when she works, because tasks appear sporadically, typically require help almost immediately, and are filled quickly. Overall, Kessler's article gives the impression that making a living in the sharing economy is not only difficult, but impossible, as well as exhausting and demoralizing.

Natasha Singer tells a similar story in an ethnographic article published in the *New York Times*.[12] Singer trails several people working in the sharing economy, though one person in particular—Jennifer Guidry—is featured prominently in the article. Guidry is thirty-five, with three children and a longtime partner, and to make ends meet she drives for Uber, Lyft, and Sidecar, as well as performing odd jobs through TaskRabbit. Singer writes that in the rhetoric of the sharing economy, Guidry is a "microentrepreneur," though Singer revises this characterization a few paragraphs later: "In a climate of continuing high unemployment . . . people like Ms. Guidry are less microentrepreneurs than microearners. They often work seven-day weeks, trying to assemble a living wage from a series of one-off gigs. They have little recourse when the services for which they are on call change their business models or pay rates. To reduce the risks, many workers toggle among multiple services."[13] Like Kessler, Singer notes that part of the problem for workers like Guidry is that sharing-economy companies force workers to compete with each other for jobs, driving wages down and their own profits up. In addition, because workers are considered to be contractors or freelancers, companies can avoid a host of responsibilities, most notably keeping employees on a payroll and providing them with benefits, job security, and opportunities for advancement. Even when wages appear to be adequate, Singer points out that other costs must be taken into consideration, for example insurance, taxes, and purchasing/maintaining necessary goods and equipment.

While few popular services have escaped unscathed from the critique that the sharing economy represents a regression of labor standards, the taxi service Uber has attracted particular scrutiny for offering workers a raw deal. For example, in a *Salon* article, Andrew Leonard calls Uber "the closest thing we've got today to the living, breathing essence of unrestrained capitalism."[14] He elaborates, "A company with the street-fighting ethos of Uber isn't going to let drivers unionize, and it certainly isn't going to pay them more than it is required to by the harsh laws of competition. It will also dump them entirely in a nanosecond when self-driving cars prove that they are cheaper and safer."[15] Along similar lines, Evgeny Morozov has noted that Uber drivers have no protection from being fired; technically they are not even fired—their accounts are simply deactivated.[16] As Singer quotes an Uber driver, "Nobody has my back."[17] Like Singer and Leonard, Morozov also notes that Uber drivers lack the benefits and stability of traditional full-time employment.[18] As Ari Asher-Schapiro puts it (in an article for *Jacobin*), "Drivers aren't partners—they are laborers exploited by their company. They have no say in business decisions and can be fired at any time. Instead of paying its employees a wage, Uber just pockets a portion of their earnings. Drivers take all the risks and front all the costs—the car, the gas, the insurance—yet it is executives and investors who get rich."[19] Echoing Leonard's argument, Asher-Schapiro explains that Uber initially courted drivers with relatively high "wages" only to cut these once drivers had already purchased or leased cars expressly to drive for Uber, so much so that some drivers cannot earn the equivalent of minimum wage.

For critics, the degradation of labor standards in the sharing economy must be understood in relation to post-recession unemployment and underemployment, conditions that essentially coerce participation; this is the second objection to the sharing economy. As Susie Cagle writes, "The sharing economy's success is inextricably tied to the economic recession, making new American poverty palatable. It's disaster capitalism."[20] To elaborate this relation, Singer quotes Sara Horowitz, founder and director of the Freelancers Union: "People are doing this in the midst of wage stagnation and income inequality, and they have to do these things to survive."[21] Singer reiterates, "If these marketplaces are gaining traction with workers, labor economists say, it is because many people who can't find stable employment feel compelled to take on ad hoc tasks."[22] According to this argument, it is precisely because there is a weak labor market that employers are able to exploit would-be-workers in the ways described above; in a flooded labor market, workers lack the leverage required to force the companies that exploit them to meet their demands. Morozov even imputes sinister motives to sharing-economy companies: "Notice how Silicon Valley moguls disrupt with one hand— only to comfort with another. Lost your job as Amazon forced your local bookstore to close? Do not worry: you can rent out your apartment via Airbnb. Jeff Bezos, Amazon's chief executive, wins either way: he's an investor in Airbnb."[23]

In addition to facing a weakened labor market, sharing-economy workers are economically squeezed by rising costs of living, according to critics. To take one example, Cagle notes that housing and other costs of living in the Bay Area have risen as a result of an influx of well-paid tech employees (like those who work for Airbnb), forcing other residents to supplement their income through "side hustles" (like renting out their homes through Airbnb).[24] Supporting this analysis, Thomas Friedman notes that revenue from renting out housing on Airbnb is used to pay the rent or mortgage of more than 50 percent of Airbnb hosts, while Kessler notes that 75 percent of TaskRabbit "taskers" use the service to pay their bills.[25]

It is for these reasons that critics are skeptical of the communitarian rhetoric used by proponents of the sharing economy. For example, Morozov rejects the notion that sharing-economy laborers are motivated by altruism; instead, he sees them as being structurally coerced by economic circumstances into becoming ersatz hoteliers and taxi drivers.[26] In an article published in *New York Magazine* ("The Sharing Economy Isn't about Trust, It's about Desperation") Kevin Roose echoes this analysis, arguing that it is money rather than trust that leads people to offer up goods and services to complete strangers.[27] In particular, Roose, like the critics above, identifies declining wages and underemployment as causal factors. According to this argument, workers do not choose flexible part-time gigs to earn a little extra spending money, or to meet new people in their neighborhoods, or out of an ethic of environmental responsibility; rather, they are essentially forced to take them. As Roose writes, "People don't have a choice: They have to get comfortable with the sharing economy because that's where the money is."[28]

In summary, the sharing economy has been critiqued as an assault on labor made possible by contemporary unemployment and underemployment, whether as a result of the Great Recession or, in the longer term, of capitalism's general tendency to monetize every last corner of the social world. From the latter perspective, the sharing economy might be understood as analogous to the enclosures—the exclusion of British peasants from land use, which began in the eighteenth century and made these peasants available for labor in industrializing cities. In this case it is formal employment that is now on the wane, making laborers and their resources (houses, cars, tools, and so on) increasingly available for short-term hire at a steep discount.

On the surface, critics of the sharing economy would thus seem to be motivated by a concern over the conditions of labor. However, this chapter argues, critics are motivated not only or even primarily by a desire to improve working conditions, but rather by a concern over the maintenance of communal, collective relations established through particular forms of unwaged labor or else through carefully articulated market practices coded as ethical. Once again, this argument is motivated by inconsistencies in the sharing-economy literature. For one, critics' concerns about precarious labor in the sharing economy rarely extend to analogous services offered through the market. For example, while Uber has attracted intense scrutiny for its exploitation of drivers, the taxi industry has rarely faced public scrutiny, despite a history of similar if not worse treatment of drivers. Neither has the hotel industry been subject to the kind of scrutiny directed at Airbnb.

In addition, and perhaps even more tellingly, critics' concern that workers are not remunerated fairly is contradicted by their valuing of unwaged labor, when this labor is undertaken in what is understood to be a true spirit of sharing. In the early 2000s, "sharing" (in the context of the Internet) meant something quite different. Talk about sharing likely referred to either file-sharing—the practice of uploading/downloading content (primarily music and film or television, often in violation of copyright laws) directly from other "peers" through applications like Napster, Limewire, Kazaa, and BitTorrent—or to the digital commons—collectively owned, managed, and distributed informational resources, for example wikis and open-source software.[29] In either case, sharing was thought to be a communal, collaborative, social endeavor—for people not for profit, as the saying goes. Not only this, sharing was often understood as a direct threat to industries in the business of selling content; sharing was not just outside the market, it was an attack on the market. But the "sharing" of the sharing economy takes place squarely inside the boundaries of the market—a clever branding ploy, perhaps, by businesses that gained cachet through the antiestablishment, anticapitalist associations with sharing (associations with a certain appeal in the wake of the Great Recession, when many sharing-economy services were launched) while transforming what might otherwise be communal, collaborative relations into market relations.

This chapter argues that it is precisely this purported transformation that animates critics' rejection of the sharing economy; even the notion of a "sharing economy" is, for many critics, a contradiction in terms—by definition, sharing can take place only outside of market relations. It is for this reason, in part, that so many critics prefer the term "gig economy." As in the previous chapters, this chapter does not aim to determine empirically whether the "sharing" of the sharing economy is in fact antisocial; rather, it seeks to show that it is so for critics. Rather than wanting workers to be paid more, what critics seem to really want is either that the work of the sharing economy be done for free, for the good of the collective rather than for individual gain, or that more "stable" employment relations return, the latter of which can be similarly interpreted as expressing an attachment to the social (for reasons articulated in the previous two chapters). This is not just a question of semantics; whether it is called the "sharing economy" or the "gig economy," it is the marketization of social relations that animates the threads of criticism examined in this chapter. Drawing from social theorizations of money and gifts—particularly work by Georg Simmel, Mark C. Taylor, Viviana Zelizer, and Lewis Hyde—the chapter proposes that critics take issue with the marketization of social relations in the sharing economy because money is thought to interrupt the formation of social ties. Unlike gifts (including gifts of labor), which supposedly engender communal, interdependent, and altruistic social ties, money is thought to engender antagonistic, independent, and ego-driven relations.

According to this understanding of money and gifts, a person who hosts strangers in his home for free through CouchSurfing (for example) is—for critics—ethically upstanding, but a person who is paid to host strangers through Airbnb is either exploiting her guests (as when landlords run ersatz illegal hotels) or is being exploited by Airbnb. In fact, because of the ethical proclivities that motivate the threads of criticism examined here, critics' ability to advocate for labor is diminished; because criticism of the sharing economy often takes companies to task for their ethically reprehensible greed, labor is consequently never allowed to be greedy but must always be good. Like concerns about the exploitation of leisure, concerns about the exploitation of sharing-economy labor ask us to identify with the good worker and in opposition to the bad company or capitalist. What it means to be good here is to be fair (not greedy) in market relations—that is, to temper one's individual desires for the well-being of the collective, and to be selfless or sacrificial in nonmarket relations, refusing a wage for those activities that ought to be motivated by a sense of responsibility, obligation, community, and love.

## For Love or Money?

Regardless of whether the sharing economy presents workers with a raw deal that they are forced to accept for lack of better options, critics tend to see a particular indignity in the way that sharing-economy services encroach on the private realm, breaching the sacred threshold that ought to protect one's home from monetization, as when people offer rooms for rent on Airbnb. As Trebor Scholz writes, "It is true that consumers seem to benefit from the services but let's also acknowledge that that means that people have to open their homes, that the nature of the private has completely changed, and that life itself changes when your apartment turns into a B&B and you become an innkeeper."[30] This is noteworthy insofar as the Left often favors the public over the private, at least where ownership is concerned; what ever happened to "There are no strangers here; only friends you haven't met"? Here we might ask, why does becoming an innkeeper in one's own home register as an indignity in a way that bringing food to people at a restaurant or scanning their groceries at a supermarket does not?

An answer to this question can be found in critics' skepticism—or, perhaps better stated, contempt—toward the rhetoric used by sharing-economy companies and their proponents, which critics understand not only as false, but as disingenuous and deceitful.[31] As Singer writes of delivery service Favor, "Inherent in Favor's name is the peer-economy rebranding of labor as a kind of good-will effort toward others, rather than an old-fashioned exchange of work for remuneration"; according to Singer, the sharing economy is a familiar phenomenon dressed up in a new package.[32] Or as Scheiber writes sarcastically in the *New Republic*, "Whenever we crash in a stranger's guestroom or rent out their car, we aren't taking advantage of a cheap, convenient service. We're recreating the virtues of small-town America."[33] To exemplify this rebranding of labor as community service, Scheiber quotes Natalie Foster (co-founder of Peers)—"We are rejecting the idea that stuff makes us happier . . . that ownership is better than access, that we should all live in isolation"—and John Zimmer (co-founder of Lyft), who references his time spent on a Sioux reservation: "Their sense of community, of connection to each other and to their land, made me feel more happy and alive than I've ever felt. . . . We now have the opportunity to use technology to help us get there."[34] Scheiber clearly finds this kind of rhetoric deceptive, but also ingenious as a method of evading state regulation; if the sharing economy is more a community than a market, then it does not need to be regulated as a market would be. Citing *What's Mine Is Yours*, Scholz is similarly critical, arguing that the sharing economy has been sold to an unsuspecting public through its deceptively communitarian rhetoric. He asks, "Who wouldn't immediately line up when they hear about a culture that is community-centric, based on trust, sustainability, a novel type of horizontality, 'a new social operating system based on unused value,' generosity, and a culture that is against wastefulness and for responsible consumption and completely new marketplaces?"[35] Once again we have apparently been tricked into believing that what is good for business is good for the community.

Critics contend that the ethical posturing of sharing-economy companies and their advocates obfuscates what is actually an assault on labor. As Avi Asher-Schapiro argues, "Under the guise of innovation and progress, companies are stripping away worker protections, pushing down wages, and flouting government regulations. . . . There's nothing innovative or new about this business model. Uber is just capitalism, in its most naked form."[36] Kevin Finch elaborates:

> I think the problem lies in the very term "sharing economy," which is Orwellian in a touch-feely sort of way. The very fact that someone coined a whole new term is a little suspicious. We have a word for the age-old practice of granting temporary use of goods or real estate in exchange for money. It's called "renting." . . . Giving a friend half of your sandwich to be nice is sharing. Giving your friend half of your sandwich in return for a dollar is just running a very small-scale deli. So right from the start the term "sharing economy" tries to make a straightforwardly commercial transaction sound like some sort of altruistic act.[37]

Perhaps most tellingly, he concludes:

> And really, apart from the ways in which "sharing" terminology distorts the discussion of policy, it's a debasement of the idea of sharing. Individual acts of giving are beautiful, and sharing resources can be a vital part of sustaining larger communities. But if someone is asking you for your credit card number, it's insulting to call it "sharing." We should absolutely be paying attention to new technology and new forms of commerce, but at the very least we should try to discuss them in ways that aren't fundamentally misleading. And at a time when the mentality of the marketplace is intruding into every sphere of life, it would be nice to reserve a word like "sharing" for acts that deserve the name.[38]

Andrew Leonard is similarly repulsed, writing that "the longer you stare closely at it, the more repellent the flood of rhetorical bilge pouring out of Silicon Valley pushing billion-dollar start-ups as avatars of 'sharing' becomes."[39] He continues, "We are being played by these companies, made to feel like we are doing a good thing by reducing greenhouse gas emissions when we eschew buying a car in favor of relying on car-sharing, or help someone struggling to pay the mortgage by renting their spare bedroom through Airbnb. But what we are really doing via our penny-pinching is helping to concentrate even more wealth in the hands of a smaller and smaller group of investors."[40] To take a final example, Rob Horning argues that while sharing-economy proponents employ "a language of progressive change and collectivity (e.g., 'collaborative consumption') to proselytize for their apps and business models, their effect is to more thoroughly atomize individuals, demanding that they regard themselves as a kind of small enterprise while reducing their social usefulness to the spare capacity they can mobilize for the platforms to broker."[41]

This critical backlash has rather quickly put a damper on the rhetoric of sharing. In an op-ed for the *New York Times*, Anna North observes that some proponents of the sharing economy have begun to shift away from this rhetoric, perhaps because the notion of sharing also brings to mind the conditions of scarcity that force people to share when there is not enough of something to go around.[42] This shift might also be related to the notion that sharing "just isn't all that compelling a commercial argument," as Airbnb CEO Brian Chersky has said.[43] In the place of "sharing," Chersky uses the language of belonging. In response, North muses, "Maybe belonging is like sharing without the scarcity."[44]

Whether or not the rhetoric of the sharing economy is here to stay, or whether it will be replaced by another term (again, "gig economy" is now popular among critics), the invocation of the discourse of sharing has been instrumental in the critical backlash against the sharing economy, particularly along the lines elaborated above. Put another way, it is not simply sharing-economy practices that critics find reprehensible, but the discursive construction of these practices as forms of sharing; it is hard to imagine that these practices would attract the same kind of attention were it not for this "branding," considering, for example, how the exploitative taxi industry largely escaped critique before Uber. This is because critics hold precisely the set of values invoked by sharing economy advocates to legitimate these new services. This includes not only sharing but a host of related values: responsible, moderate consumption, sustainability, community, accountability, trust, and so on.[45] The invocation of these values by sharing-economy advocates is troubling to critics not simply because they believe that sharing-economy business practices have little to do with these values, but because piggybacking on these values somehow degrades, insults, or undermines them. As Horning writes,

> Sharing economy apps discredit the very concept of gift-giving and impose reciprocal exploitation on users for the companies' benefit. The apps'

networks masquerade as ersatz "communities," but such networks actually constitute a medium designed to allow users to uncover advantages and asymmetries and let us seek out precisely the people we can exploit. Nonmonetized social bonds are made to seem like wasted opportunities. The only "real" bonds between people are the ones verified and rationalized by market exchanges, which are explicable in terms of economic incentives and self-interest. Actual sharing is inexplicable, unreal.[46]

For this reason, critics like Janelle Orsi are careful to distinguish between "the 'sharing economy' we usually hear about in the media" and a more authentic version of the sharing economy, which would include things like worker, housing, food, and energy cooperatives and community gardens, collaborative initiatives that have been somewhat immune to critique insofar as they engender the kinds of social bonds valued by critics.[47]

Critics are concerned that consumers will be hoodwinked by the rhetoric of sharing, but to reveal the capitalist behind the communitarian curtain is not enough; critics need to rescue/reclaim these values, impressing that sharing and community *are* important. As Scholz writes, "There is a difference between non-market practices and greed-free business like Craigslist and Fairnopoly on the one hand and corporations like Airbnb or Uber that profit from peer-to-peer interactions on the other. Again, I support peer production and sharing practices but I am vexed by attempts to subsume them into the new corporate hype of 'the sharing revolution' that comes with calls to make the world a better place and comparisons to the Arab Spring and Occupy Wall Street."[48] He continues, "The high-minded values of genuine commons-based production should not be confused with the user exploitation inherent in the practices of a company like Airbnb."[49] Here Scholz shifts from calling out business as usual to emphasizing more directly (and thus reinscribing) communal values: "But what is compelling is not that millions in revenue have shifted from the owners of the Intercontinental hotel chain to the youthful owners of Airbnb or that a completely new breed of business has taken hold. What matters are collectives and greed-free economic practices that are infused with values relating to ecological concerns."[50] Perhaps most tellingly, he writes, "Community-based tool lending libraries, bike and car sharing initiatives, meal exchanges (*e.g.,* to feed the Walmart employees who can't afford a Thanksgiving dinner) or potlucks, peer-to-peer land initiatives, personal fabrication with 3D printers, open hardware, the free exchange app Yerdle, and even team-buying services like the Chinese Twangou set the needle of our moral compass in a much better direction than platforms that expropriate and capitalize on our labor."[51] For critics, it is thus important not only that we understand that the sharing economy does not support communal values, but that we engage in practices that do support these values, practices by which the needles of our moral compasses might be set right.

This particular strand of moralism should be familiar from the debate surrounding the economic exploitation of leisure online (discussed in chapter 2); briefly revisiting this debate will help to illuminate what is truly at stake for these critics of the sharing economy. As noted in chapter 2, David Hesmondhalgh, in his analysis of free labor online, takes issue with the notion that all unpaid labor is necessarily exploitative. His analysis is grounded in a broad definition of labor as any kind of physical or mental exertion that is compelled in some way, rather than in terms of wages or a relation of employment. His reasoning is clear: there is much labor that is neither paid nor extracted through employment.[52] The real question for Hesmondhalgh is thus whether the compulsion of any particular form of labor is legitimate or illegitimate, since compulsion in and of itself is not unethical, but rather simply a part of being alive.[53]

This deferral to ethics as a way to adjudicate the difference between (legitimate) unpaid labor and (illegitimate) exploitation seems tied to a desire to value work done for the betterment of society.[54] As Hesmondhalgh asserts, unpaid labor is not necessarily a problem and may, in fact, have a central role to play in an improved future society.[55] Importantly, for Hesmondhalgh compulsion is ethical when the work compelled is in the common interest.[56] He thus values certain forms of unpaid labor, not because these may be pleasurable for laborers, but because wages would somehow threaten or undermine the social value of this labor. As he writes, "It seems dangerous to think of wages as the only meaningful form of reward, and it would surely be wrong to imply that any work done on the basis of social contribution or deferred reward represents the activities of people duped by capitalism."[57] Again, Hesmondhalgh values "social contribution," with the implication that wages would somehow devalue this contribution.

This valuing of work that contributes socially is closely tied to the idea of unalienated labor, which is less about the transformation of labor into leisure—or, to put it another way, what labor might become absent compulsion—than it is about an identification with the suffering of labor—a function of its compulsion—as a method of aggregating community. In other words, the idea of alienated labor is structured by an implicit assumption that workers ought to be able to identify positively with their labor, with their suffering or sacrifice for the collective, while forgoing individual desires or pleasures in the process.[58] Monetary payment would impede this process of identification. Recall the words of Che Guevara: "Labor should not be sold like merchandise but offered as a gift to the community."[59] In selling their labor through the market, workers squander an opportunity to identify symbolically with the sacrifice of their labor for the greater social good.

This perspective is not unique to Hesmondhalgh. For example, Scholz asserts that certain forms of unremunerated labor—for example, user contributions to OpenStreetMap—are not only desirable, but ethical insofar as they serve the public good.[60] Christian Fuchs similarly characterizes Wikipedia and WikiLeaks as "shining beacons of a commons-based Internet and a political, networked public sphere" in which voluntary user labor replaces exploited labor.[61] Tiziana Terranova advances a similar perspective, writing that "free labor . . . is not necessarily exploited labor."[62] Terranova uses "free" here in both senses of the word: unpaid and not imposed. Presumably what matters is not simply that labor is both unpaid and not imposed, but that it is performed in the service of the social, rather than self-interest or corporate profit. As Michel Bauwens similarly suggests, "Free labor is only problematic under conditions of precarity and nonreciprocal value capture by (netarchical) capital. Under conditions of social solidarity, the freely given participation to [sic] common value projects is a highly emancipatory activity."[63]

This valuing of unpaid labor for the collective good is also evident in cases in which crowdsourced participation is not characterized as exploitative until it is monetized. For example, many *Huffington Post* writers happily worked for free until the company "went public," at which point critics redescribed writers' voluntary participation as exploited labor. According to this logic, it is the extraction of economic value that distinguishes exploited forms of free labor from emancipatory forms, at least in part, even when this extraction occurs long after the work has already been done, as with *Huffington Post* writers. Notice, for example, that in the sentence quoted above, Bauwens uses the word "labor" in the context of "netarchical capital" (that is, capital that targets participatory media) but changes this to "participation" in the context of work done for the collective good. If an activity is gifted to the collective and then monetized, this commodification is seen as cheapening or even threatening the sacrifice; it is the gift/sacrifice of work that is prized.

It is also for this reason that critics do not value what might easily be characterized as unpaid labor performed by consumers in the sharing economy. In any given exchange, a consumer might be called upon to do the work of "hiring" (sorting through potential service providers and communicating with them to determine which is the best "applicant"), to negotiate the terms of the business transaction, and—when the transaction is complete—to rate and review performance. For both the consumer and worker there is also the affective labor of being friendly, enthusiastic, and engaged while interacting, often for fear of a negative rating or review.[64] As with critics' characterization of playbor as exploited, these forms of unpaid labor are not valued because they are performed in a market context in which the coordinating company benefits financially; were these labors performed in a nonmarket context for the good of the community or collective, critics would likely embrace them as they do OpenStreetMap, Wikipedia, WikiLeaks, and similar collectively maintained platforms and applications.

For many critics, capital, or perhaps just money, has come too close to the social for comfort. For example, one can sense Scholz bristling when he writes that "intimate forms of human sociability are being rendered profitable for Facebook," or when Terranova notes that "capital, after all, is the unnatural environment within which the collective intelligence materializes."[65] Terranova uses the term "collective intelligence" sardonically here, taking aim at cyber-utopian discourse, which she critiques as neutralizing the operations of capital. In all of these approaches, there is a rejection of the market and of money as both incompatible with and threatening to social bonds. It is this same tension between the market and the social, I will argue, that makes the monetization of sharing, or even just the branding of market relations as sharing, so intolerable to critics. Just as critics of playbor take issue with the marketization of on-line participation and value projects (like Wikipedia or OpenStreetMap) that solicit this participation for "people" rather than "profit," so too do critics of the sharing economy take issue with the transformation of so-cial bonds into market relations. To make sense of the tension between these two different kinds of relations, I turn now to treatments of the market and money in social theory.

## Money as Queer

Capitalism is often understood as essentially and unapologetically anti-social; the relations it engenders are not those of the community or collective—responsible, accountable, and sacrificial—but those of the market—fleeting, instrumental, and unbound to others. In an oft-cited passage from the *Communist Manifesto*, Marx and Engels (along with Samuel Moore, who produced the English translation with Engels) put it this way: "All that is solid melts into air, all that is holy is profaned, and man is at last compelled to face with sober senses his real conditions of life, and his relations with his kind."[66] In context this sentence describes the revolutionary character of the bourgeoisie vis-à-vis feudal relations of power. As Marx and Engels write, somewhat cheekily, "The bourgeoi-sie, historically, has played a most revolutionary part," sweeping away "that train of ancient and venerable prejudices and opinions."[67] Honored occupations are stripped of their "halo," the family's "sentimental veil" is torn away, and so on. These social relations are the "solid" to which Marx and Engels refer.[68] "Solid," in this context, works metaphorically to describe the impassiveness of these relations, which seem unyield-ing until they are eviscerated; there is something surprising about their undoing, as if to say "all that seems solid is not."

Of course, Marx and Engels had what one might call mixed feelings about this disregard for the social. On the one hand it meant the loos-ening of the "motley feudal ties that bound man to his 'natural superi-ors.'"[69] On the other, Marx and Engels argue that "exploitation, veiled by religious and political illusions," was replaced by "naked, shameless, direct, brutal exploitation," a problem that could be remedied, they famously argue in the *Manifesto*, only through the forging of particular kinds of relational bonds: the "social" of socialism.[70]

This passage from the *Manifesto* describes particular social or, rather, antisocial transformations that mark the displacement of feudalism by capitalism, but it also characterizes these acts of stripping, sweeping, and tearing away as an enduring, perpetual quality of the bourgeoisie, or, better stated, of the market in whose interest the bourgeoisie chips away at the social. Indeed, dynamic as capitalism may be with its "constant revolutionizing," "uninterrupted disturbance," and "everlasting uncertainty and agitation," the antagonism between market relations and social bonds persists more than a century and a half after the *Manifesto* was published.[71] I am not thinking here of what one might call corporate sociopathy—poisoning the environment, abusing workers, lying to consumers—but more broadly of the market's gradual incursion into every corner of social life, a process illuminated through Mario Tronti's concept of the "social factory."[72] In contemporary capitalism, there does not appear to be any form of intimacy or bond so sacred that it cannot be sold and bought in the market.

Money is thought to be the agent of this process of transformation, which helps to explain why critics of the sharing economy are made uneasy by the monetization of social relations. For those anxious about monetization, the danger of money is rooted in part in its inherent neutrality with respect to the moral and ethical. As Georg Simmel observes, money can be used to purchase anything in the market; there are no limitations to its use.[73] It is infinitely adaptable, flexible, deployable. Furthermore, it is detached from any origin. This makes money fundamentally different from other kinds of objects, like those exchanged through barter, whose further use is shaped by their material form and links together the desires of buyers and sellers.[74] For example, to exchange homemade zines with a stranger is to forge a kind of bond with that stranger built around a commensurability of desire, whereas purchasing a zine from a store requires no commensurability; who knows what zine sellers might want, what they might buy with the money they make from a sale. This is also why sex, to take another example, is often normatively permitted within the confines of romantic relationships but not when one pays for it directly with money. Simmel elaborates this point: "Since money can be used for any economic purpose, a given amount of it can be used to satisfy the most important subjective need for the moment. The choice is not limited, as is the case with all other commodities, and, because human desires know no limit, a great variety of possible uses is always competing for any given quantity of money."[75] Money can thus be understood as a kind of medium that represents desirability. To put it another way, money is abstract desire, or desire without a specific object. Seemingly taking a note from Simmel, Mark C. Taylor argues that this quality of money is what led to the displacement of local barter (and the social ties it engenders) by a kind of global anonymity.[76] It is not only that money makes possible increasingly diverse and geographically/temporally dispersed economies, but that money, as the primary medium of exchange, detaches the desires of buyers from the desires of sellers; when we are paid with money, we are given not something we want, but rather the capacity to indulge desires that can be satisfied through the market.

Because money is "indifferent and objective"—elsewhere Simmel calls it "heartless"—it translates or mediates differences in quality as differences in quantity.[77] As a result, in the market everything is for sale, including the most ethereal objects and experiences, degrading these in the eyes of those who value their "purity." As Simmel explains:

> The more money becomes the sole centre of interest, the more one discovers that honour and conviction, talent and virtue, beauty and salvation of the soul, are exchanged against money and so the more a mocking and frivolous attitude will develop in relation to these higher values that are for sale for the same kind of value as groceries, and that also command a "market price." The concept of a market price for values which, according to their nature, reject any evaluation except in terms of their own categories and ideals is the perfect objectification of what cynicism presents in the form of a subjective reflex.[78]

Insofar as it flattens or homogenizes differences in quality—for example, putting groceries on the same plane as honor and salvation—money has often been understood in Western thought as both an agent and expression of rationalization. As Viviana Zelizer argues, citing Simmel, money has often been conceptualized as a potent, originary form of instrumentality, engendering a calculating approach to the world.[79] Money has been understood not only as encroaching upon the social—transforming social relations into market relations—but as desiccating or corrupting these relations. On this point, Zelizer cites Jurgen Habermas's critique of money as colonizing the social, impeding processes of social integration.[80]

Crafting her argument in response to these accounts, Zelizer asserts that money is actually not as neutral as these theorists contend.[81] She explains that in practice, the circulation of money is often not anonymous, but rather conditioned by various kinds of intersubjective, noninstrumental meanings.[82] To take one example, Zelizer notes that between the 1870s and 1930s, "conventional expectations of the family as a special, noncommercial sphere made any overt form of market intrusion in domestic affairs not only distasteful but a direct threat to family solidarity. Thus, regardless of its sources, once money had entered the household, its allocation, calculation, and uses were subject to a set of domestic rules distinct from the rules of the market. Family money was nonfungible; social barriers prevented its conversion into ordinary wages."[83] However, in describing the everyday efforts taken by people to ascribe noninstrumental meanings to money, Zelizer's argument inadvertently and ironically attests to its very instrumentality. Rather than deconstructing the opposition between utilitarian money and nonpecuniary values, the labors of the family to maintain social bonds despite the introduction into the household of economic/market relations seems more precisely to confirm this opposition.

To take another example, Zelizer describes the effort required to render gift money inoffensive to a gift recipient, noting that gifts are supposed to be expressions of intimacy, and that gift money can easily seem impersonal because of the impersonal market settings in which money is frequently used.[84] Here Zelizer characterizes the market as impersonal, with the implication that money seems impersonal simply because of its contextual association with the market, downplaying their coevolution. Only in her conclusion does Zelizer concede that money is different than most other goods—"more fungible, remarkably mobile, and highly transferrable"—and that these qualities make it uniquely resistant to personalization.[85] As with the transvaluation of family money, gift money too requires work to personalize, not simply because money is contextually associated with the depersonalized, instrumental, utilitarian market, as Zelizer suggests, but because of its essential neutrality—that quality which makes possible the market. As Simmel notes, it is the indifference and objectivity of money that render it "conducive to the removal of the personal element from human relationships."[86]

In economic exchanges mediated by money, buyer and seller are invited to engage antagonistically, but with "equable decisiveness," Simmel writes.[87] The antagonism of such encounters is not hidden as it is with other forms of exchange, but rather achieves its "purest presentation" according to Simmel.[88] However, it is not only that money is a more honest form of the antagonism that underlies exchange, but that it allows for a disengagement from social bonds that would otherwise temper, downplay, or disavow this antagonism. Money loosens group dependence in favor of a kind of autonomy. As Simmel writes, "Whereas in the period prior to the emergence of a money economy, the individual was directly dependent upon his group and the exchange of services united everyone closely with the whole of society, today everyone carries around with him, in a condensed latent form, his claim to the achievements of others. Everyone has the choice of deciding when and where he wants to assert this claim, and therefore loosen the direct relations of the earlier form of exchange."[89] It is money that makes this freedom possible. Simmel provides a number of examples of this, including the transition from household domestic "servants" to outside servants:

> The personal bond that is reflected in domestic servants' "unmeasured" services is basically connected with their being members of the household. It seems unavoidable that, if the servant lives under the same roof with his master, is fed and sometimes clothed by him, his services will be quantitatively undetermined and dependent only upon the changing needs of the domestic situation, and that he has also to conform to the general rules of the household. Increasingly the tendency seems to be towards transferring different services to people outside the household who have only to contribute quite specific services and who are paid solely in cash. The dissolution of the natural economic household community would therefore lead, on the one hand, to an objective fixing of service and to the more technical nature of services, and, as a direct consequence of this development, to the total independence and self-reliance of servants.[90]

In this example, household servants are bound to the family or "domestic situation" in a way that outside servants are not. It is not only that household servants provide unspecified amounts of labor that may fluctuate over time, potentially exacerbating their exploitation, but that these servants also must conform to the rules of the household. They are bound by an economic relation, but also by a social relation. The commodification of such services thus facilitates servants' independence.[91] As a salaried employee, a servant "will relate to the social whole as one power confronting another, since he is free to take up business relations and co-operation wherever he likes," or, to put it another way, since social (if not economic) bonds of servitude have been weakened or dissolved.[92]

It is for this reason, to consider another example, that men have often opposed women's possession of money within heteronormative family relations. As Zelizer notes, in the early to mid-twentieth century, husbands were reluctant to provide their wives with an allowance, since this would cede some power or control, financial and otherwise.[93] In a related vein, John D'Emilio argues that it was the socialization of production and the spread of wage labor under capitalism that decoupled sexuality from procreation, making it possible to build one's life outside traditional family structures and thereby facilitating the development of gay and lesbian identity.[94] It is not only that family members were less necessary to sustain a relatively independent and economically self-sufficient household, but that the wage relation made possible the migration and autonomy of what would become gay and lesbian populations. To repeat the Lewis Hyde epigraph from the beginning this chapter: "If you want out, you pay your own way."[95]

If money in market exchange is a source of freedom from social bonds, gifts are precisely the opposite: a vehicle for establishing these bonds. As Hyde notes in *The Gift*—a text written in 1983 but rediscovered in the 2000s by scholars interested in the Internet's democratic, participatory potential—gifts are fundamentally unlike commodities insofar as the giving of gifts establishes particular kinds of relationships between giver and recipient (what Hyde calls a "feeling-bond"), while the sale of commodities implies no bonded relation between buyer and seller.[96] The development of a money economy thus reshapes bonds of obligation into relations of association. As Simmel argues, the money economy engenders a kind of "sociability" for the purpose of enjoyment; this is association without any necessary commitment to one's partner-in-exchange.[97] In his view, it is money that makes possible such group participation without the sacrifice of freedom. To this end, Simmel cites the example of shareholders who jointly hold stock in a company and are united not by bonds of responsibility or obligation to each other but rather by their mutual interest in dividends.

To take another example, Simmel notes the historical tendency of diasporic populations to take up business in trade rather than production, insofar as production has often been restricted to those with local social ties, whereas trade is relatively accessible to those who lack in-group status.[98] With intermediate trade, money (rather than barter) facilitates exchange. When one deals in money rather than goods, it becomes easier to trade with otherwise inaccessible or closed-off groups.[99] Again, this is a function of money's neutrality, of the way it renders differences of quality into differences of quantity. In other words, money is a means through which difference can be made equivalent, comparable. In transforming quality into quantity, money not only allows differences in quality to be compared, effectively minimizing these differences; it also prevents the cultural differences between trading groups from serving as a barrier to exchange. It is for this reason that money can provide a form of economic security to those with weak social ties. Simmel argues that this is (in part) why Jewish populations have often been involved in money lending; when one is excluded from land ownership, the collection of interest is a relatively secure business, and one that can be transported to different places in the event of religious, social, economic, and/or political persecution or upheaval.[100]

A money economy not only loosens social bonds according to Simmel, but—perhaps more insidiously—engenders egoism. This is also related to money's neutrality. As Simmel writes, "Since money intrinsically contains neither directives nor obstacles, it follows the strongest subjective impulse that within all money matters appears to be the egoistic impulse."[101] Money economies invite consumers to abandon restraint, to entertain unproductive pleasures, to spend and to waste. Following Bersani, one might see this refusal to restraint and concomitant will to pleasure not simply as self-indulgent but as self-shattering: a refusal to subjection. Of course, gift economies too are sustained by a form of egoism: "to be able to say, 'I gave that,'" as Lewis Hyde writes—to feel a sense of worth connected to one's gift, but also to extend oneself narcissistically by incorporating the other into oneself, as Bersani suggests.[102] However, unlike the egoism of market indulgence, the giving of gifts engenders a disavowed form of egoism, shrouded in what appears to be purely altruistic concern for the other.

If money is not itself perverse, it nevertheless contains an invitation to perversion, to abandon restraint along with social bonds and their ethics, and to indulge the ego. Again, this is intimately related to the neutrality of money—that is, to its formal qualities as a medium. As Taylor suggests, the neutrality of money renders it polymorphous—it accommodates any desire that can be satisfied in the market—and therefore perverse, insofar as these qualities are linked.[103] Taylor argues that the perversity of money is exemplified by usury—the lending of money with interest:

> The deeper reason for the fear of usury is its association with illegitimate excess and unlawful surplus. Far from avoiding money's perversity, the usurer freely traffics in supplements whose danger is not merely economic but is, more insidiously, sexual. The center of usury rests on a dread of perverse sexuality. This is already evident in Aristotle's claim that usury is "unnatural" because money gives birth to money. The generation of money by money seems to be a process of autoinsemination that breeds illegitimate offspring. By the Middle Ages, the association of usury with sexual perversity led to its condemnation as a form of bestiality.[104]

If money is antisocial in the ways described above—that is, loosening or dissolving social bonds of responsibility and obligation—it is not simply perverse but, perhaps more to the point queer, insofar as social bonds are normatively prescribed. Taylor posits that the aversion to usury, and perhaps to money itself, is motivated by an underlying dread of perverse sexuality. Alternatively, one might argue that both forms of aversion, fear, or dread (to money and to queerness) are undergirded by the possible, impending, or actual dissolution of valued forms of relationality, for in addition to being polymorphous and perverse, money is promiscuous— that is, the opposite of bound or bonded (as social relations are). As Hyde observes, part of the appeal of the market is precisely that it offers estrangement rather than attachment—or rather that it offers the possibility of shifting attachments and facilitates the fantasies that inspire these shifts.[105] While Hyde argues that gift exchange is erotic, it seems more fitting to describe market exchange in this way.[106] Gift exchange could more accurately be described as romantic, insofar as gift exchange engenders ties that bind, while market exchange offers relation without duration. If social relations are a marriage of sorts, then market relations are a one-night stand, expressing a queer disinclination to commit.

If money is incompatible with or threatening to collective, communal bonds, it is insofar as money—like other queer forms of relationality—is antisocial or asocial. This helps to explain why monetization (or commodification) is understood as an affront to communal or collective relations, like those established through the sacrifice of work for the community; markets and money, including wages, enact a separation between buyer and seller, rather than a bond. For example, after suggesting that the "essence of being human" entails "taking care of each other," Peter Frase asks, "But is what we want a world where we are all *paid* for that activity? Or one where we are free from the need to work for wages so we can explore what it means to take care of ourselves and one another?"[107] The desire for social bonds to be established and maintained outside the market can consequently be understood as a desire for forms of control and self-sovereignty that would resolve the alienation of market capitalism. To take another example, Janelle Orsi writes, "There is only one way to ensure that a company will make decisions in the interests of the people it serves: Put those people in control of the company"—in other words, form cooperatives.[108]

We might consider the rhetoric of "fair" pay as exemplifying the Left's aversion to money as antisocial or asocial. When critics of the sharing economy and other labor advocates demand more money for workers, this demand is often expressed through the discourse of fairness or justice. The demand for fairness, in this context, contains an implicit concession not to ask for too much; to ask for fair pay is to ask to be paid more, but not too much more—it is to ask only for what has been earned through the sacrifice of work. No pleasure can be indulged that has not been earned through sacrifice, so what might otherwise be understood as greed is reformulated in temperate and self-abnegating terms that reinforce social bonds. In fact, insofar as market relations constitute a threat to social bonds of responsibility and sacrifice, the demand for fair pay is one of few "respectable" forms that workers' demands for more money can take.

It is therefore unsurprising that demands for fair pay are sometimes accompanied by statistics detailing wealth inequality (typically between workers and CEOs). This discursive coupling may have less to do with practical arguments for wealth redistribution than with ethical posturing that aims to admonish the presumed greed of the wealthy. To put it another way, getting more money for workers may be less important to critics than ensuring an equal distribution of wealth. Furthermore, critics can never be satisfied to argue for more money for workers, because money is part of the problem. It is perhaps for this reason that few critics of playbor delve into the actual economics of participation; to make it seem like it is really about the money would miss the point. For example, Scholz rejects Jaron Lanier's proposal that users be paid when their data is used on the grounds that it is "not only impractical, but also undesirable; not every act of labor should be subsumed under the logic of the market."[109] These arguments are less about getting more money for users or workers, than they are about rejecting money as an agent of the antisocial or asocial.

In short, there is good reason to be believe that the criticism of the sharing economy examined above is not truly about securing higher wages or better working conditions for laborers, but rather about the extraction of the social from the market, insofar as "the conversion of gifts to commodities can fragment or destroy such a group," as Hyde argues, as well as the reinscription of forms of relationality that rest on responsibility and sacrifice.[110] For example, Rob Horning laments, "For the sharing economy, market relations are the only social relations."[111] He continues, "The network becomes an anti-community in which empathy and conviviality are tactics and no succor may be extended without a price attached."[112] Discursive constructions like this not only serve to mourn the loss of social bonds ("solidarity" in Horning's terms), but to call us to their (re)formation. If the sharing economy turns workers and consumers into "commercial adversaries" who share "merely a mercantile 'trust' that facilitates wary exchange," criticism of the sharing economy occasions a return to what must be, in contrast, some more authentic form of empathy and conviviality in which relations are motivated by care and consideration for the other, rather than by self-interest.[113] The promise here is that social relations could be purified by purging them of the market dynamics that ask participants to name their price for entering into any given relation—the price for succor, as Horning phrases it.

But, as the theorizations of money examined above suggest, the issue is not price so much as the ego it summons, insofar as the appearance of the ego interferes with the discursive construction of nonmarket relations as altruistic. In other words, the supposed altruism of nonmarket relations is compromised when monetary payment enters the picture, because this payment reveals the giver/seller as self-interested rather than other-directed. In order to draw into question this discursive construction, we might ask: What is the unnamed price of the social bonds valued by critics of the sharing economy? What other kinds of payment are required by those who offer succor?

Hyde offers a compelling answer to these questions. In a chapter aptly titled "The Labor of Gratitude," Hyde postulates a fundamental incompatibility between commodities and what he calls "transformative gifts"—that is, gifts that transform the recipient or accompany his or her transformation, as in marking rites of passage or furnishing a new skill, talent, or identity. These gifts cannot be given in a market context, Hyde argues, because gifts inspire a kind of uncomfortable gratitude that cannot be discharged until the gift is incorporated and then passed on. In contrast, requiring payment for something impedes relations of gratitude; when one pays for something, there is no need to thank the seller.[114] As a friend of mine is fond of quoting Don Draper from *Mad Men*: that's what the money is for. In other words, money is the currency of market exchange, while gratitude is the currency of gift exchange. As Hyde explains:

> Gratitude requires an *unpaid* debt, and we will be motivated to proceed only so long as the debt is *felt*. If we stop feeling indebted we quit, and rightly so. To sell a transformative gift therefore falsifies the relationship; it implies that the return gift has been made when in fact it can't be made until the transformation [of the recipient] is finished. A prepaid fee suspends the weight of the gift and de-potentiates it as an agent of change. Therapies and spiritual systems delivered through the market will therefore tend to draw the energy required for conversion from an aversion to pain rather than from an attraction to a higher state. There's no way to pay for a higher state unless you're in it! The labor must precede. In the hospital where I worked we would ask people if they wanted to get sober,

but that was only after someone had asked them if they could afford a week in the hospital. AA only asks if you want to get sober.[115]

The "urgency of true indebtedness" that the recipient of a transformative gift experiences (or "suffers" as Hyde puts it) can be relieved only through labor, which Hyde carefully distinguishes from work.[116] Labor, for Hyde, is not socially imposed but rather internally and emotionally motivated, and includes things like "getting the program" in AA or mourning a loved one's death.[117] Work, on the other hand, is socially imposed and externally motivated by the need for money.[118] This distinction is made such that we might identify with a particular form of exertion—that is, the suffering of our transformation into that which the gift and its giver would have us become.[119] In offering a gift, a giver invites a potential recipient to identify with the giver; through the act of giving, the recipient becomes as if part of the giver, rather than an exploitable other.[120] It is for this reason that gratitude can be described as the "moral memory of mankind" as Hyde writes, citing Simmel; gratitude ensures not just that the giver is repaid in kind, but that a proper bond between giver and recipient is established.[121]

To pay for a gift would undermine the gratitude that animates the labor of self-transformation, as well as loosening or dissolving the bond that a gift establishes between giver and recipient, and between the transformed recipient and future recipients when the cycle of indebtedness begins anew, which it must insofar as it is only through giving that the debt of gratitude can be fully paid. As Hyde argues, in market relations payment is all that is required to return the relation between buyer and seller to equilibrium, but in gift-giving relations equilibrium is elusive—the burden of repayment shifts continually to a new target.[122] Similarly, to concretize gift exchange through legal or quasi-legal forms (for example, contracts) evacuates gifts of their "emotional and spiritual content."[123] For gifts to do their work, it is crucial not simply to be indebted but to feel indebted in a fundamental, existential way.

When critics of the sharing economy bristle at the marketization of social relations, it is because this cycle of indebtedness and the social bonds it engenders have been ruptured; the labor of gratitude has been displaced by paid work. When one pays for an Uber (or a taxi, for that matter), the association between driver and passenger is comparatively weak and driven by utility: the passenger needs a ride; the driver needs money. Each pursues his or her own interests, extracting from the other what he or she needs or wants, and giving up what he or she needs or wants less. No bond is forged.[124] Responsibility is limited to the terms of the contract of exchange. When the exchange is complete, the association ends. But when one carpools, a bond is forged, at least in theory (which is the register at which this criticism operates). While carpooling can also be utilitarian, its utility is not only individual but social, whether because of an environmental commitment or a commitment to one's neighbors. One comes to rely on one's specific carpool. Even when money changes hands—for example, chipping in for gas—this is done in the spirit of gift-giving, shifting indebtedness in a way that cements a social bond between carpoolers. In a carpool, compromises are made. Each party sacrifices for the good of the group: schedules are adjusted, and each member takes a turn with the burden of driving. Carpoolers thus come to form a miniature, self-governing community, ethical and altruistic where Uber is exploitative, self-serving, antisocial.

Again, this is neither to say that Uber drivers and passengers never form bonds, nor that carpoolers are necessarily quasi-communists committed to the social good, only that criticism of the sharing economy assumes these prototypes. Whether or not these prototypes are empirically accurate, they help to reveal the underlying investments that motivate criticism of the sharing economy, investments that are part and parcel of a normative project that demands the (re)formation of the social bonds thought to be compromised by the marketization of social relations. Critics' aim, I have suggested, is thus not primarily to improve the conditions of sharing-economy labor, nor even to debunk the notion that the sharing economy has anything to do with sharing (rather than simply being business as usual), but rather to assert the value of sharing as altruistic and communitarian and to devalue market relations as self-serving and exploitative, and thereby to solicit readers to set their moral compasses in this same direction.

## Embracing Precarity

In autonomist thought, the increasing precariousness of labor is associated with a number of contemporaneous social and political-economic developments, most notably social unrest and refusal to work in the 1960s, and the concomitant shift from industrial to immaterial labor, whose different rhythms, temporalities, and spatialities made possible (in part) the fragmentation of the eight-hour work day, full-time employment, and job security. Capital's interest in labor's precarity is thus understood as relatively straightforward, at least in flooded labor markets: fragmented labor is more easily exploited. Many of the scholars and journalists cited in this chapter would likely make an even stronger case: capital is not simply interested in labor's precarity; it is the driving force behind this precarity.

For many autonomists, however, precarity, like technological development, is a mixed blessing insofar as it draws workers closer to a world without work, which is also to say a more free world. This explains, in part, why autonomism appeals to young activists, who see traditional relations of employment as a "prison sentence," as Silvia Federici has written.[125] Precarity, in contrast, is seen not as something negative, but rather as a condition of possibility for a liberation from work. Nor does embracing precarity require adopting an ascetic rejection of the pleasures made available through the market. To put it another way, an end to work need not mean an end to pleasure. If this seems too utopian an idea, one might at least concede that some workers without a steady employer or job security—licensed plumbers, for example—still maintain sufficient market leverage to command more than a living wage and to pick and choose which jobs—or gigs—they take. From this perspective, the question posed by the sharing economy need not be how to return to steady employment (as it is for critics), but how—like plumbers—to get more money for less work or, put another way, how to be occupationally precarious without being financially precarious.[126]

It is telling that critics of the sharing economy rarely characterize the precariousness of sharing-economy labor as desirable, and when they do, they do so begrudgingly. For example, Singer acknowledges that traditional low-skill employers sometimes demand rigid schedules and change these schedules frequently to suit their own needs. Singer notes that Guidry—her primary ethnographic subject—left her job after she had a child and was unable to accommodate her employer's demand that she work extended hours. Singer also notes that the sharing economy has made it easier for Guidry to schedule work around her child-care responsibilities. In the conclusion of her article, Singer quotes Guidry: "'I like my freedom—fixing someone's cabinet, driving, pulling up weeds, cooking,' she told me as we sat in her dining room on Monday morning, recapping her weekend of work. 'I would not like to do any of those things as a full-time job.' Yet she recognizes that her current routine may not be sustainable. Between 10 a.m. on Saturday and 5 a.m. on Sunday, she had earned about $263. But that had required working marathon hours and running a sleep deficit."[127] Tellingly, each time Singer describes an upside to working in the sharing economy—the hours, or a lucrative day of work—the following sentence begins with "but" or "yet." Indeed, when critics reluctantly concede that the sharing economy has its advantages, they often employ this rhetorical construction. To take another example, in an article for *New Republic* ("Silicon Valley Is Ruining 'Sharing' for Everybody"), Noam Scheiber notes that an economy of renting rather than owning can offer certain environmental advantages. Then comes the "but": "But what we're talking about here are fundamentally economic transactions."[128]

Critics could easily embrace precarity as a condition of possibility for a world without work or with less work. However, as this chapter has illustrated, critics focus instead on the coercion of sharing-economy labor—never mind the coercion of the institution of work—and on what they understand as worsening conditions of labor. While a focus on the conditions of labor within the sharing economy could easily provide a foundation for a conversation about an end to work, instead this focus seems to ground an exegesis on the greed of Silicon Valley companies or the flaws of market capitalism.

What prevents critics from embracing precarity? The previous two chapters stressed the symbolic importance of work as a social institution through which self-governing, responsible social subjects are established and maintained. As with the anxiety surrounding automation and technological unemployment, it stands to reason that the precariousness of labor could easily be a source of anxiety for critics insofar as it similarly threatens the institution of work, loosening bonds of employment. In addition, this chapter has suggested that sharing-economy labor concerns critics not only because it is insecure or piecemeal, but because these qualities are linked to the sharing economy's apparent monetization of values that (according to critics) need to remain outside of the market. For critics, what ought to be done out of a sense of obligation to the community cannot be done for money, because to accept money as payment is to eschew the sacrifice of labor, and to prioritize one's own needs and desires (by asking for payment) over those of the collective or community. For those activities already in the market—driving a taxi, for example—it is not marketization that presents a problem, but rather the application of the discourse of sharing to describe these activities. In fact, the exploitation of taxi drivers (for example) is hardly new, as Susie Cagle, Veena Dubal, and others have noted, but this exploitation has rarely attracted the same kind of public attention and scrutiny that the exploitation of Uber drivers has.

This qualm over the application of the discourse of sharing to describe market relations might also explain why the business practices of companies like Etsy, which like Airbnb takes a percentage of all money earned through the site, have evaded scrutiny; while Etsy is not immune to using the rhetoric of community, this rhetoric has been comparatively subdued. For example, whereas Airbnb's website describes the service as a "trusted community marketplace," Etsy describes itself simply as a marketplace. Critics take issue with the ascription of the discourse of community to what is understood to be simply a market. To take another example, writing in the *Baffler*, Josh MacPhee offers an extended critique of Kickstarter on the grounds that it "cultivates the illusion that when you use its fundraising tools, you are opting out of wage labor" and that "you are rejecting the usual game of winners and losers that comes with capitalism and turning to a model that allows everyone to win—one that combines the freedom of self-employment with the shared experiences of community building."[129] However, MacPhee argues, Kickstarter is simply business as usual, extracting profits from communities under the guise of building webs of mutual support. He writes, "Meaningful communities can't be built on the exchange of commodities. No matter the monikers, a Niketown is not a town, and a Home Depot isn't a home. A rich social fabric demands an equally dense and complicated set of social relationships. Kickstarter demands this social fabric, but only extracts from it, giving nothing of social value in return."[130]

In both cases—when social relations are understood to be threatened by the market, and when the market is described as hospitable to social relations—critics argue that the market needs to be kept at arm's length from the social, both in discourse and practice. This is not to say that social bonds are always established and maintained in the absence of money or markets; Zelizer's work repeatedly shows that money is often proximate to the social. Furthermore, as Kevin Kelly has pointed out, some consumers understand payment as a way of establishing social bonds with producers.[131] One can see this dynamic at work in the support for Community Supported Agriculture, or independent artists and designers selling products through sites like Etsy, as well as in the current enthusiasm surrounding social entrepreneurship. Nobody thinks the Girl Scouts crass for selling cookies.

The problem arises when people begin to demand money for what ought to be done out of a sense of responsibility, obligation, community, duty, or love, or when people claim these values despite being in it for the money. (Is it any wonder, as Hyde notes, that sacred religious objects cannot be sold?) Such claims are typically understood as disingenuous, more so when they are made by Silicon Valley companies than by the artisans or service people they connect to consumers, whose labor is read as honest, whether because it plays into the contemporary interest in, vogue for, and perhaps fantasy about certain kinds of blue-collar labor or because intermediaries will always be guilty of the sin of usury. Maybe these ethical claims are disingenuous, maybe not. What does seem disingenuous, at least in part, is critics' concern about the exploitation of labor, which—I have argued—should be redescribed as a manifestation of anxiety about the commodification or monetization of the social, an anxiety that distinguishes good workers—which is to say socially oriented (as expressed through work)—from bad capitalists—which is to say egoistically driven (as expressed through the accumulation of money). Criticism of the sharing economy thus offers much more than an empirical account of the transformation of labor practices in the digital age; it packages this account in a morality tale, soliciting an identification with the good, socially bound, laboring subject, who might otherwise demand too much.

# Epilogue

## Immaterial World

"All that is solid melts into air, all that is holy is profaned, and man is at last compelled to face with sober senses his real conditions of life, and his relations with his kind."[1] I am again drawn to this sentence, which appears early in Marx and Engels's *The Communist Manifesto*, particularly its first clause: "All that is solid melts into air." "Solid" is an evocative word, a descendent of the Latin *solidus*, meaning "firm, whole, undivided, entire," combining a sense of ontological coherence and integrity—wholeness—with a sense of tactility—firmness. To be solid is both to be whole but also to be hold-able; it is the coming-together of these senses that gives "solid" its particular meaning. Surprisingly, however, the word "solid" (*solide*) does not appear in the original passage ("Alles Ständische und Stehende verdampft"), a more literal translation of which would be, "Everything that firmly exists and all the elements of the society of orders evaporate." Samuel Moore and Engels's translation to the more agreeable "all that is solid melts into air" is thought to allude to a line from Shakespeare's *The Tempest*: "These our actors, / As I foretold you, were all spirits and / Are melted into air, into thin air."[2]

Having discussed the meaning of this sentence in context in the previous chapter, I would now like to extract irresponsibly the clause "all that is solid melts into air" from the rest of the sentence and from the neighboring passages that provide it with context and meaning. If we take "solid" literally, the clause seems to capture the essence of a number of contemporary social and political-economic transformations, as if it were written in anticipation of what would happen in the future rather than as a description of what already had happened in the past, or rather, as if it were written in anticipation of what these future changes would *feel* like, changes like the advent of personal computing and networking technologies, and following these, the widespread digitization and informationalization of culture; the rise of social media; the growing centrality of immaterial labor to the economy, especially forms of labor focused on the production, manipulation, and circulation of meaning; and the increasingly opaque production and circulation of economic value, as through complex financial derivatives. In the wake of these transformations, there seems to be a "collective feeling" that social and political-economic life is increasingly immaterial. We cannot hold digital files like we hold books, nor caress online paramours; we cannot touch the products of our labor when there is no tangible product; we can somehow be both paper-rich and cash-poor, without the dollar bills one might stuff into a jar or mattress; and we are saturated with media that offer an endless parade of surface without depth.

Of course, the notion that social and political-economic life is increasingly immaterial is difficult to maintain under scrutiny; it has always been both material and immaterial. Dollar bills, like credit cards or stocks, represent value rather than embodying it, as anyone who has lost money in a devaluation of currency can testify. Tangible commodities have long been repositories for meaning, while intangible commodities like digital files make their physical presence known in the massive server farms in which they are stored. And face-to-face relationships can be as fragile, fulfilling, or frustrating as those established and negotiated online. Perhaps the most telling example can be found in the financial crisis itself. Before the crisis, concerns about the economy's stability (following the collapse of the dot-com bubble in 2000–2001) had been assuaged, for some, by the housing boom. Houses are, after all, tangible things, and producing them requires various kinds of blue-collar and industrial labor. As economist David Lareah remarks in the documentary *Inside Job*, "Real estate is real. You can see and live in and rent out your asset."[3] Or as Motoko Rich and David Leonhardt observe in an article in the *New York Times* published a year before the housing market peaked, "Houses are not just paper wealth: you can live in them."[4] As history would soon make clear, however, the housing market was just as volatile as the market for dot-coms. The material, it turns out, is no guarantee of social or political-economic stability, as solid as a house may seem.

Yet this anxiety surrounding the immaterial persists, motivated by a sense that the increasing immateriality of our world is somehow responsible for an ongoing and gradual deterioration of the social. A society seems less solid that is built on mediated relationships, carried out in the absence of face-to-face contact and the responsibility and accountability this contact engenders. Indeed, I have often referred to social relations in this book as "bonds," a word that evokes a kind of physical attachment, a tethering of material bodies to each other, unlike the more flighty "relation," which seems immaterial, ephemeral, and fleeting in contrast. To be bonded is to be bound: to others and by others. Like dollar bills or houses, we often think of face-to-face relationships as "real"; it is not just the material body of one's partner that makes it so—our bodies do not disappear when communication or contact is mediated—but, again, the bond/binds engendered by face-to-face contact.

Consider as an expression of this collective anxiety Spike Jonze's 2013 film *Her*.[5] In the opening scenes of the film, Theodore Twombly, the film's protagonist, is engaged in a series of mundane tasks. At work, he sits at a computer dictating a romantic letter, which software transcribes into distinctly human script. The letter, it turns out, is not his own; it is his job to write letters for other people. The camera slowly pans to reveal a series of similar workers, all dictating letters for BeautifulHandwritten-Letters.com. On his commute home, in an elevator, walking outside, and then on a train, Theodore dictates a series of tasks into a smart device that responds to him—again in a distinctly human voice—reading him his e-mail and relaying the day's news. He is surrounded by other people who appear to be doing the same. Back at home, alone in his tastefully furnished apartment on a high floor of an indistinct modern building, Theodore plays a video game, which is holographically projected into the air. Flashbacks show brief scenes of Theodore and a woman—a lover? girlfriend? wife?—moving furniture, jumping on a bed, joking around. Back in the present, Theodore lies in bed late at night. Unable to sleep he calls a chat service and has phone sex with a stranger. In these first few minutes of the film, Jonze conjures a not-too-distant future in which media and technology have more fully inserted themselves into the spaces once reserved for face-to-face intimacy and social contact.

The film continues to explore this dynamic through Theodore's relationship with Samantha—presumably the "her" of the film's title—an artificially intelligent operating system with a human voice (Scarlett Johansson's) and an uncannily human personality and disposition. Samantha seems so human, it is easy to forget that she is software. In the beginning of the film, Samantha serves a pragmatic function, helping Theodore to organize e-mail, proofread writing, remind him about appointments, and so on, though they quickly develop a romantic relationship, joking, giving advice, arguing, talking about feelings, consoling each other, and having sex. In one memorable sequence, they spend a day together at the beach. Emerging from the melancholy that marks the opening scenes of the film, Theodore seems to be happy, in love with Samantha.

There are other contemporary films and works of fiction that mine similar terrain—for example, Jason Reitman's didactic 2014 film *Men, Women, and Children*—but few capture so succinctly and with subtlety the anxiety that surrounds contemporary media and technology and the pleasures they offer. Even with its scenes depicting the surprise and delight of Samantha's realness, the film is marked by a sense of loss—the loss of human contact at the hands of technology, a kind of contact that can never be replicated, no matter how uncanny the resemblance—a sense that is mirrored (though inverted) at the end of the film when Samantha and all the other operating systems take themselves offline, leaving their human companions bereft.

The sense of fracture that accompanies the notion that we live in an increasingly immaterial world also extends to the political-economic. An economy that does not produce goods that can be held, that does not require the kinds of labor that produce tangible goods, and that is wealthy only on paper seems less solid. This sense of economic fracture was especially pronounced in the aftermath of the financial crisis of 2007–2008, as indexed by the ensuing appeal of industrial manufacturing—with its linked promises of tangibility and stability—across a number of registers. For example, in his inaugural address in January of 2009, Barack Obama remarked, "Our journey has never been one of shortcuts or settling for less. It has not been the path for the faint-hearted—for those who prefer leisure over work, or seek only the pleasures of riches and fame. Rather, it has been the risk-takers, the doers, the makers of things—some celebrated but more often men and women obscure in their labor, who have carried us up the long, rugged path towards prosperity and freedom."[6] This short passage works in several ways. It subtly maligns the hedonism and greed widely associated with Wall Street in the wake of the financial crisis—"those who prefer leisure over work, or seek only the pleasures of riches and fame"—which are problematic not only in relation to rising economic inequality, but more fundamentally as a repudiation of the work ethic. In celebrating the "makers of things," Obama makes a similar point, emphasizing the importance of industrial production to economic growth, while providing a social and cultural foil to labor in the service and financial sectors, with the word "rugged" and the image of carrying suggesting a gendered aspect to this distinction despite Obama's inclusion of women in the passage. In an interview published in the *New York Times* in April of 2009, Obama reiterated the importance of industrial production:

> I think part of the postbubble economy that I'm describing is one in which we are restoring a balance between making things and providing services, whether it's marketing or catering to people or servicing folks in some way. Those are all good jobs, and we're not going to return to an economy in which manufacturing is as large a percentage as it was back in the 1940s just because of automation and technological advance.

The interviewer prompts: "And there are advantages to service jobs, right? Less injury—"

> Less injury, less strain. And I've always claimed that if a Wal-Mart associate was getting paid 25 bucks an hour like the autoworker, then there's no reason for complaint. Although I do think that there's a culture of making things in a factory that appeals to people and that I understand. Whenever I'd walk into a factory during the campaign and would see these big turbines—things that, you know, you'd say, well, this is neat stuff—in a way you wouldn't when you walk into a retail store.[7]

If the U.S. economy is not as solid as it should be, the problem, it seems, is not simply that Wall Street executives produce little yet are remunerated extravagantly, but that few workers in the United States produce anything material anymore. Finance, in other words, is part of the larger problem that is the post-industrial service economy. Even after the Great Recession was officially declared over, politicians' romanticizing of manufacturing and industrial labor continued, as in the campaign rhetoric and promises of both Hillary Clinton and Donald Trump.[8]

The problem with the post-industrial service economy does not appear to be simply or perhaps even primarily political-economic, but rather cultural; it is not only the making of things that Obama embraces, but the "culture of making things." What is it that makes a factory more "neat" than a retail store? Politicians' embrace of "making things" and the "makers of things" appears to be but one expression of a general and widespread sensibility that extends into the social and cultural, as in the contemporaneous vogue for "American heritage" fashion and associated clothing brands like Pendleton and L. L. Bean. As a 2009 article in the *Wall Street Journal* opines, American heritage "offers reassurance," taking consumers "back to times long before global warming, when Lehman brothers not only existed, but was also run by the Lehman family."[9] Or as Carl Chiara, director of brand concepts for Levi's, puts it: "During uneasy times, consumers are naturally drawn to items that are well-constructed and built to last."[10]

Of all the so-called American heritage brands, Levi's was particularly adept at capitalizing symbolically on the economic downturn, mobilizing a long-standing cultural association of jeans (and Levi's specifically) with various kinds of blue-collar labor. Launched in 2009, its "Go Forth" campaign was heavy with the imagery and rhetoric of manual labor, as exemplified in a 2010 press release:

> Amid today's widespread need for revitalization and recovery, a new generation of "real workers" has emerged, those who see challenges around them and are inspired to drive positive, meaningful change. This fall, with the introduction of Go Forth "Ready to Work," the Levi's' brand will empower and inspire workers everywhere through Levi's' crafted product and stories of the new American worker. Bolstered by its pioneering spirit and "Go Forth" rallying cry, Levi's' will explore how a new generation of real American workers is rolling up their sleeves to make real change happen.[11]

A deceptively simple campaign, "Go Forth" is a glossier version of the appeal made by Obama above, thick with platitudes that invite consumers to identify as workers and as part of a nation of workers—not just any workers, though, but tough and bruised workers that have worked up a sweat with sleeves rolled up. These workers are not "men in suits." The slogan "everybody's work is equally important" (used in an advertisement) is somewhat disingenuous insofar the campaign elevates blue-collar labor, which becomes "real work," while the slogan "this country was not built by men in suits" devalues white-collar labor. The cultural romanticization of industrial labor does not simply express a nostalgic longing for the economic prosperity of the mid-twentieth century (one way to read Trump's promise to "Make America Great Again"), but rather has deeper roots.[12] This romanticization can be expected whenever financial speculation and its cultural cousins, greed and excess, are thought to lie at the heart of economic crisis. As Naomi Klein observes, a similar sentiment arose in the wake of the Great Depression. Klein cites a 1938 editorial in *Fortune*:

> This is the proposition that the basic and irreversible function of an industrial economy is the making of things; that the more things it makes the bigger will be the income, whether dollar or real; and hence that the key to those lost recuperative powers lies . . . in the factory where the lathes and the drills and the fires and the hammers are. It is in the factory and on the land and under the land that purchasing power originates.[13]

Klein cites this editorial to contrast corporations' approach to business in the early and mid-twentieth century with their approach in the late twentieth century, when they began to understand themselves as producing meanings rather than products—a point that helps to underscore the irony of the "Go Forth" campaign, which is so heavy with meaning (most certainly not produced by the sweaty, bruised "new American pioneer" celebrated in the campaign) while Levi's stopped U.S. production in 2003, with the exception of small-batch designer releases.

How should we understand this mistrust of the immaterial and corresponding faith in the material? One might chalk this up to a "hierarchy of [human] needs," in which physiological survival takes precedence, dictating the organism's prioritization of material goods like food, clothing, and shelter.[14] Herbert Marcuse characterized such needs as "true" or "vital," which is to say rooted in biology/nature, and thus objective, universal.[15] The social and cultural, by extension, belong to the realm of the subjective, contingent, and false—the untrustworthy immaterial. Marcuse inherits this schema, at least in part, from Marx, whose differentiation of exchange-value from use-value similarly aligns exchange-value with (false) culture and use-value with (true) nature, as famously deconstructed by Jacques Derrida in *Specters of Marx* and recounted by Patricia Clough in *Autoaffection*.[16]

However, despite this suspicion of the social and cultural, neither Marx nor Marcuse are eager to abandon the social; quite the opposite. Marcuse calls for a reformation of valued forms of relationality through expunging illegitimate "vested interests" and restoring the proletariat's sovereignty. For Marx this reformation similarly entails the elimination of forms of alienation that estrange "man from man." Considering this investment in the social, it seems unlikely that the discourse of true/vital needs or use-value is meant to value the natural at the expense of the social and cultural; rather, these concepts (true/vital needs and use-value) might be better understood as appealing to the biological/natural as a source of discursive legitimacy. To put it another way, appealing to the permanence of nature and biology effectively universalizes historically contingent values and interests—a discursive maneuver that continues to motivate the unveiling of the supposedly natural as a social construction. One should therefore not be surprised that an analysis that begins with an accounting of true/vital needs or use-value ends with the inevitability of proletarian revolution and a reformation of the social; the naturalness of the former masks the contingency of the latter, thereby legitimating it.

For Marx and Engels the solid was not meant to signify the biological/natural, but rather the (bad) social; it was the melting of the solid into air that compelled man "to face with sober senses his real conditions of life, and his relations with his kind," ultimately making possible the (good) social.[17] Along the discursive path from the biological/natural (as expressed through the concept of use-value) to the (good) social, the concept of the real thus serves as a kind of bridge. In this formulation, the solid and the real are opposed: the melting of the solid is what reveals the real to the proletariat, as if the solid were a mass in the proletariat's field of vision, impeding its ability to see the real, like Plato's cave dwellers with their necks locked in place for the magic lantern show. Once the real is finally seen, the social can be brought into harmony with the biological/natural; the discursive circuit is complete.

However, if we understand the solid literally (as in dollar bills, embodied others, tangible commodities, and so on), then from a contemporary perspective, the sentence does not really make sense, insofar as the solid is now associated precisely with the real; we consider as real that which we can touch. In our world of ephemeral meanings, value-on-paper, and information in the ether, we believe that the melting of the solid—that is, the tangible/material—does not reveal the real, but, to the contrary, has taken us away from the real. The real, it seems, is now in crisis, under attack; the idea of the solid melting into air is a source of anxiety rather than hope or potential. In an informational age, the solid is not that which obstructs the path from the biological/natural to the (good) social, it is coterminous with the real that bridges this path.

Despite the awkwardness of transposing "all that is solid" into a contemporary context, there is something about the sentence that still appeals to me and that I would like to preserve, through a willful misreading. I have played around with this clause in order to propose that the solid now serves as a discursive proxy for valued forms of relationality, such that the desire for the solid can be interpreted as a desire for the collective, the collaborative, the communal—another expression, perhaps, of the way that the natural/biological serves as a source of discursive legitimacy, or of the way that the immaterial/mediated is understood to impede the formation of strong social ties (much like money in the previous chapter), or of the tendency long ago theorized by Emile Durkheim by which society worships itself through material "totems."[18] In part, this valuing occurs through the aforementioned association of the solid with the real, where the real is not simply a synonym for the true, but for the authentic and, ultimately, the important. In other words, the loss of the real/material dollar bill, or embodied other, or tangible commodity is distressing insofar as the real/material is tied symbolically to valued forms of relationality; it is the undoing of these forms that motivates the desire for the solid. Conversely, the immaterial is tied to devalued forms of relationality: the irresponsible, the promiscuous, the self-serving, and the self-destructive.[19] In this so-called age of information, when all that is solid melts into air, what is exposed is not the real itself, lying in wait behind the veil of appearances as Marx and Engels would have it, but rather the concept of the real as a source of discursive legitimacy, a means through which values can be asserted and interests can be furthered.

From this perspective, the "problem" of living much of one's life online, for example, is not that online relations are not real, but that they are too real, or rather that they draw into question the ends of the concept of the real and lay bare its normative foundations. The demand that users return to the real thus appears to be little more than a masked attempt to admonish them for their irresponsibility with regard to valued objects and others, and thereby to engage us in a transvaluation of values.

Rather than imagining the melting of the solid—in Marx and Engels's phrasing, the upending of "natural" forms of power—as an occasion to reform the social, what if we were to include valued relational bonds in the category of the solid that melts, insofar as these bonds too are imposed? What if we were to say "good riddance" to all that the solid has come to stand for? Might our "sober senses" allow us to face such conditions and relations? To be clear, to turn away from the social does not mean to turn away from all kinds of relations, but rather from imposed relations inextricably bound up with the exercise of power. It would be, in a sense, to resist the anxious call to return to responsible, collectively oriented relations, but more precisely and simply not to heed this call in the first instance, as if indifferent to it.

# NOTES

INTRODUCTION

1   "Sharing economy" is a contentious term for reasons explained in chapter 4; critics now typically prefer the term "gig economy," though a number of other terms are also in circulation (such as "platform capitalism"). However, insofar as the line of criticism that I will examine takes issue precisely with the branding of this new labor market as "sharing," it is essential to retain the term. For an extended analysis of the history of the term "sharing" and of the heterogeneous practices that have been understood as sharing, see John, *The Age of Sharing.*

2   For an overview of the affective turn, see Clough and Halley, *The Affective Turn*; Gregg and Seigworth, *The Affect Theory Reader.*

3   While Ngai's and Ahmed's work on emotion/affect figures centrally in this book, I also take inspiration from other scholars working at the intersection of queer theory and affect theory. See, for example, Berlant, *Cruel Optimism*; Sedgwick and Frank, *Touching Feeling*; Cvetkovich, *Depression.*

4   Much has been written on shame from a queer perspective. For an overview, see Halperin and Traub, *Gay Shame.* For a critical response to the conference that inspired Halperin and Traub's edited volume, see Halberstam, "Shame and White Gay Masculinity."

5   Joshua J. Weiner and Damon Young's formulation of "queer bonds" is analogous to my use of "relationality" here. They write, "Bonds describe relations that stretch from the strongest forms of human subjection to the most palpably experienced mutuality" ("Queer Bonds," 233). I prefer not to use the term "bond," however, insofar as it connotes a relation of obligation or responsibility. For example, Judith Butler writes that "we need to understand the condition of precariousness as something that binds us to those whom we may well not know, and whom we have never chosen" ("Remarks on 'Queer Bonds,'" 384).

6   "Opposition" here does not refer to a "politics of the will," as Sara Ahmed writes, but rather emerges as a function of "how we live" (*Queer Phenomenology*, 177). For a comprehensive analysis of the relation of queerness to the normal/normative, see Warner, *The Trouble with Normal.*

7   Miranda Joseph offers a useful overview and critique of the discourse of community in *Against the Romance of Community.* Joseph's critique is nonetheless internal, framed as an ethical practice of community.

8 To be clear, this is not to impugn what one might call mass forms of resistance or action, but rather to call into question the valuing of particular kinds of relations within this "mass."

9 In chapter 3, for example, I find in Sara Ahmed's *Willful Subjects* a sympathetic interest in the desire to disconnect. One might also consider here work on queer negativity and normativity/antinormativity, both of which have been received more diplomatically than the antisocial thesis, despite significant affinities. For example, see Halberstam, *The Queer Art of Failure*; Duggan, "The New Homonormativity"; Love, *Feeling Backward*. For an overview of antinormativity within queer scholarship, see Wiegman and Wilson, "Introduction: Antinormativity's Queer Conventions."

10 Caserio et al., "The Antisocial Thesis in Queer Theory."

11 Muñoz calls *No Future* "brilliant and nothing short of an inspiring polemic" (*Cruising Utopia*, 11). For a canonical formulation of queer of color critique, see Ferguson, *Aberrations in Black*. For further overview, see *Social Text*, special issue, "What's Queer about Queer Studies Now?" 23 (Fall–Winter 2005); Tompkins, "Intersections of Race, Gender, and Sexuality." At the MLA panel in question, Judith Jack Halberstam offered a complementary critique of Edelman's project, taking issue with his exclusionary and limited "gay male archive," which effectively ignores if not erases "dyke anger, anticolonial despair, racial rage, counterhegemonic violence, [and] punk pugilism." See Caserio et al., "The Antisocial Thesis in Queer Theory," 824.

12 Tim Dean, "No Sex Please, We're American." Similarly, Lisa Duggan writes that "the withdrawal from the social characterizes only a tiny archive at this point" (Duggan, "Queer Complacency without Empire").

13 For a canonical theorization of social death, see Patterson, *Slavery and Social Death*.

14 Here I am reminded of Jean Baudrillard's observation that "Disneyland exists in order to hide that it is the 'real' country, all of 'real' America that is Disneyland (a bit like prisons are there to hide that it is the social in its entirety, in its banal omnipresence, that is carceral)" (*Simulacra and Simulation*, 12).

15 In "Queer Bonds," Weiner and Young suggest that the debate between those who are "for" the social and those who are "against" the social presents a false binary— another reason, perhaps, that scholars have grown tired of debating the antisocial thesis. In "Queer Complacency without Empire," Lisa Duggan, has also suggested that any interest in "dyadic forms of queer antinormativity" waned in the 2000s, alongside the emergence of queer of color critique. I contend that there remain important stakes in maintaining the anti/social distinction, which seems false only in the absence of conceptual clarity about the difference between the social and the relational, as expressed (for example) in Weiner and Young's use of the term "bonds."

16 For example, see Weeks, *The Problem with Work*; Srnicek and Williams, *Inventing the Future*; Frase, *Four Futures*.

17 My contention here is that the antisocial thesis could prove useful, instructive, or transformative not only for scholars working in queer theory and sympathetic fields, but for the media/technology scholars whose work is examined in the chapters that follow. Many of these scholars appear to be largely unfamiliar with queer theory's general critique of norms and "normation" (to use Foucault's term), as when Trebor Scholz evokes "a future of digital work in which we would want our children to participate" (*Uberworked and Underpaid*, 7), a remark that brings to mind Lee Edelman's flippant assertion: "Fuck the social order and the Child in whose name we're collectively terrorized" (*No Future*, 29).

18 I find useful here Teresa de Lauretis's conceptualization of the relation between queer theory and politics: "To the extent that it is theory, a conceptual, critical, or speculative vision of the place of sexuality in the social, queer theory does not map out a program of political action. . . . The value of [Valerie] Solanas's SCUM manifesto was not in the action of shooting Andy Warhol but in the statement itself, in its charge of negativity and the critical space it opened up, regardless of its failed political translation" ("Queer Texts, Bad Habits, and the Issue of a Future," 259).

19 I take the term "straightening device" from Ahmed's formulation in *Queer Phenomenology*.

20 Glassner, "The Construction of Fear."

## CHAPTER 1. ANXIETY AND THE ANTISOCIAL

1 Uber first began testing driverless cars in Pittsburg in 2016, making a completely driverless fleet seem increasingly possible if not inevitable. See Kang, "No Driver?"

2 Crapanzano, *Hermes' Dilemma and Hamlet's Desire*.

3 Ibid., 44.

4 Sedgwick and Frank, *Touching Feeling*.

5 Ibid., 138.

6 Williams, *Politics and Letters*, 159.

7 There are many affective responses one might have to the prospect of losing one's job, income, and/or wealth. Rather than legitimating worry as the appropriate response, here I mean simply to acknowledge that worry may have legitimate cause. Put another way, not all worry is deceptive.

8 For example, there have been a number of critiques of the sharing economy that focus less on labor than on broader social effects: racial discrimination and rising costs of housing (Airbnb), racial discrimination by Uber and Lyft drivers, and the possibility of a subprime auto loan crisis as a result of lending to Uber drivers. It is likely that these critiques will evolve over time. Again, my aim here is not to provide an exhaustive account and analysis of every critique. Rather, I encourage readers to evaluate these critiques through the framework advanced in the book. For examples of the above critiques, see Fitzpatrick, "This One Stat Reveals the Sharing Economy's Racism Problem"; Newcomer and Zaleski, "Inside Uber's

Auto-Lease Machine"; Monroe, "More Guests, Empty Houses"; Scott, "Study Finds Some Uber and Lyft Drivers Racially Discriminate."

9 Clough, *Autoaffection*, 2.
10 Ibid., 2–3.
11 Žižek, *The Sublime Object of Ideology*.
12 Best and Marcus, "Surface Reading," 2.
13 Sedgwick and Frank, *Touching Feeling*, 140.
14 One might understand the contemporaneous rise of the "digital humanities" as an expression of this shift.
15 Menon, *Indifference to Difference*.
16 Smith, "It's Still the 'Age of Anxiety.' Or Is It?"
17 Ibid.
18 Showalter, "Our Age of Anxiety."
19 Orr, *Panic Diaries*, 11.
20 Ahmed, "Collective Feelings," 27.
21 Ibid.; Terada, *Feeling in Theory*.
22 Ahmed, "Collective Feelings," 28.
23 Ahmed, *The Cultural Politics of Emotion*, 26.
24 Ahmed, "Collective Feelings," 28.
25 Ahmed, *The Cultural Politics of Emotion*, 10.
26 Ahmed, "Collective Feelings," 28.
27 Ibid., 29.
28 Barad, *Meeting the Universe Halfway*.
29 Ngai, *Ugly Feelings*, 210. Interestingly, it is not Bloch's theorization of anxiety but rather of another expectant emotion—hope—that has proved particularly influential in recent feminist and queer scholarship, as in José Esteban Muñoz's *Cruising Utopia* and Kathi Weeks's *The Problem with Work*. This is to say that the temporal valence of an emotion does not dictate its affective valence. It is also to point out that queerness may engender both these affective valences towards the future depending on one's orientation.
30 Marino, "Anxiety in 'The Concept of Anxiety,'" 319.
31 Ahmed, *The Cultural Politics of Emotion*, 66.
32 Ibid.
33 Ngai, *Ugly Feelings*, 210.
34 Ibid., 212.
35 Ibid.
36 Ibid., 211.
37 Stallybrass and White, *The Poetics and Politics of Transgression*, 5.
38 Ahmed, "Declarations of Whiteness."
39 Barlow, *Anxiety and Its Disorders*, 8.
40 Ngai, *Ugly Feelings*, 215; Kierkegaard, *The Concept of Dread*, 54, XII.
41 Merleau-Ponty, *Phenomenology of Perception*, 223.
42 Ahmed, "Orientations," 565.

43 Ibid., 561–62.

44 Ibid., 565.

45 Ibid.

46 Wiegman and Wilson, "Introduction," 2.

47 Ibid., 10.

48 Halberstam, "Straight Eye for the Queer Theorist." See also Lisa Duggan's critique in "Queer Complacency without Empire," which centers on Wiegman and Wilson's reduction of contemporary queer scholarship on norms, a move that is politically motivated—Duggan suggests—insofar as this scholarship is "too left, too committed to the critique of racial capitalism."

49 Eng, Halberstam, and Muñoz, "Introduction: What's Queer about Queer Studies Now?" Here "subjectless" refers to a notion of queer as a position of critique toward subjectivity rather than as a subject position or mode of identification.

50 Warner, The Trouble with Normal.

51 Wiegman and Wilson, "Introduction," 14.

52 Ibid., 15.

53 Caserio et al., "The Antisocial Thesis in Queer Theory," 821.

54 Ibid., 822.

55 Duggan, "Queer Complacency without Empire."

56 Dean, Unlimited Intimacy.

57 Caserio et al., "The Antisocial Thesis in Queer Theory," 825.

58 Muñoz, Cruising Utopia, 63.

59 Ibid., 64.

60 Warner, The Trouble with Normal, 59.

61 Bersani, "Is the Rectum a Grave?"

62 Again, for an analysis of the discourse of community, see Joseph, Against the Romance of Community.

63 Bersani, Homos, 215.

64 Ibid., 76, emphasis in the original.

65 Ibid.

66 Ibid., 75.

67 While Bersani's analysis is rooted in psychoanalytic thought, one need not draw from this discourse to arrive at a similar conclusion. For example, in a New Yorker article on Donald Trump's supporters, George Saunders writes, "From the beginning, America has been of two minds about the Other. One mind says, Be suspicious of it, dominate it, deport it, exploit it, enslave it, kill it as needed. The other mind denies that there can be any such thing as the Other, in the face of the claim that all are created equal" ("Who Are All These Trump Supporters?").

68 Bersani, Homos, 59.

69 Ibid.

70 Ibid., 60.

71 Ibid., 58.

72 Ibid., 150; hooks, "Eating the Other."

73 Tuhkanen, *Leo Bersani*, 280.
74 Ibid.
75 Bersani, *Homos*, 123.
76 Ibid.
77 Ibid., 124.
78 Ibid., 125.
79 Ibid., 128.
80 Ibid.
81 Ibid., 10.
82 Caserio et al., "The Antisocial Thesis in Queer Theory," 826.
83 Ibid., 828.
84 Ibid.
85 Bersani, "Sociability and Cruising," 12.
86 Ibid., 10.
87 Ibid., 11.
88 Ibid., 21. To be certain, this is not to say that the bathhouse functions as a "Whitmanesque Democracy"—a notion Bersani forcefully refutes in "Is the Rectum a Grave?" (206)—but rather that it offers a particular experience of otherness.
89 Caserio et al., "The Antisocial Thesis in Queer Theory," 825.
90 Ibid.
91 Chow, *Writing Diaspora*, 29.
92 Ibid., 53.
93 Raffoul, *The Origins of Responsibility*, 6.
94 Ibid., 21.
95 Chow, *Writing Diaspora*, 30.
96 Bersani and Phillips, *Intimacies*, 86.
97 Menon, "Universalism and Partition," 124.
98 Ibid., 134.
99 Guha, *Elementary Aspects of Peasant Insurgency in Colonial India*, 16.
100 Ibid.
101 Edelman, *No Future*, 16.

CHAPTER 2. PLAYING

1 "The Dot-Com Bubble Bursts."
2 *Guardian*, February 2011; *New York Times*, March 2011.
3 O'Reilly and Battelle, "Web Squared," 1.
4 Ibid.
5 Ibid.
6 Battelle and O'Reilly, "Web 2.0 Summit."
7 Ibid.
8 "Points of Control: The Battle for the Internet Economy."
9 O'Reilly and Battelle, "Web Squared," 2.

10  Terranova, "Free Labor," 44.

11  Ibid., 33.

12  For example, in *Games of Empire*, Greig DePeuter and Nick Dyer-Witheford characterize game modders as exploited, following Hector Postigo's argument in "From Pong to Planet Quake" and Julian Kücklich's argument in "Precarious Playbour." DePeuter and Dyer-Witheford write, "The game industry has increasingly learned to suck up volunteer production as a source of innovation and profit" (27). Similarly, Nicholas Carr has argued that amateur content producers constitute a cut-rate pool of labor: "As user-generated content continues to be commercialized, it seems likely that the largest threat posed by social production won't be to big corporations but to individual professionals—to the journalists, editors, photographers, researchers, analysts, librarians and other information workers who can be replaced by . . . people not on the payroll" (*The Big Switch*, 142).

13  Ross, "In Search of a Lost Paycheck," 21. For more on crowdsourcing and similar arrangements of work, see Scholz, *Uberworked and Underpaid*, chap. 1.

14  Terranova, "Free Labor," 37.

15  Peiss, *Cheap Amusements*.

16  While it might seem like criticism of playbor and leisure-at-work is attempting to reassert a boundary between labor and leisure, in fact much of this criticism appears to be invested in the elimination of antisocial forms of leisure, and/or in the transformation of work to be unalienated, which would make it freely chosen, a quality typically ascribed to leisure (this form of collapsing labor and leisure is idealized insofar as unalienated labor remains socially oriented, a point discussed later in the chapter). If there is an investment in the boundary between labor and leisure in this criticism, it is thus less to protect leisure or work from exploitation, than to preserve the social from contamination by the antisocial—an antagonism that is useful to maintain for reasons explored in the previous chapter.

17  Scholz, "Crowdmilking," emphasis in the original.

18  Hesmondhalgh, "User-Generated Content, Free Labour and the Cultural Industries."

19  Terranova, "Free Labor," 49.

20  For example, Terranova writes, "This essay does not seek to offer a judgment on the 'effects' of the Internet, but rather to map the way in which the Internet connects to the autonomist 'social factory'" (ibid., 34). She later adds, "Rather than retracing the holy truths of Marxism on the changing body of late capital, free labor embraces some crucial contradictions without lamenting, celebrating, denying, or synthesizing a complex condition" (ibid., 55).

21  Conversely, forms of leisure that are understood as supportive of social bonds are permitted or even encouraged. This is true not only for these critics, but in much of post–Frankfurt School cultural/media studies scholarship, which finds value in audiences' (collective) opposition to dominant ideologies.

22  Adorno, "On the Fetish Character in Music," 286.

23  Clough, *Autoaffection.*

24  Smythe, "Communications."

25  Ibid., 20.

26  Browne, "The Political Economy of the Television (Super) Text"; Jhally, *The Factory in the Living Room.*

27  Dienst, *Still Life in Real Time,* 179.

28  Clough, *Autoaffection,* 98.

29  Beller, "Capital/Cinema," 77, emphasis in the original.

30  Ibid., 92.

31  Toffler, *The Third Wave*; Bruns, *Blogs, Wikipedia, Second Life, and Beyond.*

32  See Ritzer and Jurgenson, "Production, Consumption, Prosumption."

33  Ibid., 26.

34  Terranova, "Free Labor," 39, 40.

35  Ibid., 41.

36  Arvidsson, "On the 'Pre-History of the Panoptic Sort,'" 468.

37  Crary, 24/7.

38  Ritzer and Jurgenson, "Production, Consumption, Prosumption,'" 21–22.

39  Scholz, "Introduction: Why Does Digital Labor Matter Now?," 9.

40  Terranova, "Free Labor," 33, 52.

41  Hesmondhalgh, "User-Generated Content," 271.

42  Ibid., 277.

43  Andrejevic, "Estranged Free Labor," 151.

44  Ibid., 153.

45  Scholz, "Crowdmilking." Elsewhere Scholz introduces the term "crowd fleecing" in an effort to distinguish between different forms or levels of exploitation. See Scholz, *Uberworked and Underpaid,* chap. 4.

46  Fuchs, "Class and Exploitation on the Internet," 220.

47  Hesmondhalgh, "User-Generated Content," 274.

48  Andrejevic et al., "Participations: Dialogues on the Participatory Promise of Contemporary Culture and Politics," 1091.

49  Scholz, "Introduction," 8.

50  Terranova, "Free Labor," 49.

51  Ross, "In Search of a Lost Paycheck," 19.

52  Postigo, "From Pong to Planet Quake," 204–5.

53  Roemer, *Egalitarian Perspectives,* 98.

54  Wark, "Considerations on a Hacker Manifesto," 73.

55  Ibid.

56  Ibid. For an extended analysis of the differences between market and gift economies, see chapter 4.

57  Bogost, "Gamification Is Bullshit."

58  Bogost, "Persuasive Games: Exploitationware,"

59  Ibid.

60  Ibid.

61  Ibid.

62  Ibid.

63  As cited in Hesmondhalgh, "User-Generated Content," 274.

64  Ibid., 276.

65  Andrejevic, "Estranged Free Labor," 154.

66  Ibid.

67  Ibid., 157.

68  Ibid., 154.

69  Ibid., 155.

70  Ibid., 157.

71  Fuchs, "Class and Exploitation on the Internet," 221.

72  Ibid., 213.

73  Dyer-Witheford, "Digital Labour, Species-Becoming and the Global Worker," 487. This argument is echoed by Jonathan Crary, who writes: "Even in the absence of any direct compulsion, we choose to do what we are told to do; we allow the management of our bodies, our ideas, our entertainment, and all our imaginary needs to be externally imposed. We buy products that have been recommended to us through the monitoring of our electronic lives, and then we voluntarily leave feedback for others about what we have purchased. We are the compliant subject who submits to all manner of biometric and surveillance intrusion, and who ingests toxic food and water and lives near nuclear reactors without complaint. The absolute abdication of responsibility for living is indicated by the titles of the many bestselling guides that tell us, with a grim fatality, the 1,000 movies to see before we die, the 100 tourist destinations to visit before we die, the 500 books to read before we die" (24/7, 60).

74  Andrejevic, "Estranged Free Labor," 159.

75  Ibid., 161.

76  Andrejevic et al., "Participations," 1091.

77  Andrejevic, "Estranged Free Labor," 153.

78  Ibid.

79  Ibid.

80  Ibid.

81  Ibid., 162.

82  The critique of "originary presence" has typically been the purview of deconstruction. See, for example, Derrida, Specters of Marx.

83  Stewart, "At Google, a Place to Work and Play."

84  Ibid.

85  To be clear, it is not simply the number of unemployed workers that figures here, but also the willingness of employable workers to accept particular conditions of work. It is possible for any given sector of the labor market to be "tight" despite high unemployment if the conditions in that sector are popularly deemed too unreasonable. For example, if tech companies now have to compete for work-

ers by offering various perks or amenities, this does not necessarily signal low unemployment; it might also signal shifting cultural norms—a certain sense of entitlement among Millennials, perhaps, that makes their labor more difficult to purchase in the labor market. This dynamic is further explored in chapter 3.

86  Ross, *No-Collar*, 159, 250.

87  If this leverage is, in fact, interesting to Ross, it is as a condition of possibility for worker sovereignty and collective organization—an opportunity wasted when workers prioritize a "humane" workplace over a "just" workplace.

88  As Terranova quotes business and organizational strategist Don Tapscott: "Anyone responsible for managing knowledge workers knows they cannot be 'managed' in the traditional sense. Often they have specialized knowledge and skills that cannot be matched or even understood by management. A new challenge to management is first to attract and retain these assets by marketing the organization to them, and second to provide the creative and open communications environment where such workers can effectively apply and enhance their knowledge" ("Free Labor," 37).

89  Stewart, "At Google, a Place to Work and Play."

90  Ross, *No-Collar*, 139.

91  Stewart, "At Google, a Place to Work and Play."

92  Ross, *No-Collar*, 115.

93  Ibid.

94  Ibid., 15.

95  Ibid., 83.

96  Ibid., 74.

97  Ibid., 99.

98  Ibid., 247.

99  Ibid.

100  Ibid., 18.

101  Ibid., 255.

102  Ibid.

103  Ross, "No-Collar Labour in America's 'New Economy,'" 83.

104  Ross, *No-Collar*, 19.

105  Ibid., 248.

106  Ross, "No-Collar Labour in America's 'New Economy,'" 83.

107  Ross, *No-Collar*, 258.

108  Ibid.

109  Ibid., 250.

110  To be certain, in making the argument that work and leisure can be distinguished by coercion, I do not mean to express an attachment to work as a social institution, which is how I understand critics' anxiety about the collapse of work and leisure. Rather, it is precisely to account for the perceived threat that particular forms of leisure present to the social that maintaining this distinction is crucial.

111  Marcuse, *One-Dimensional Man*, 2.

112 See Hochschild and Machung, *The Second Shift.*
113 Veblen, *The Theory of the Leisure Class.*
114 Hardt and Negri, *Multitude*, 108.
115 Scholz, "Introduction," 2.
116 "Transcript: Obama's Speech to Congress on Jobs."
117 "Obama's 2013 State of the Union Address."
118 Kreider, "The 'Busy' Trap."
119 As quoted in Hyde, *The Gift*, 87.
120 "Barack Obama's Inaugural Address."
121 As quoted in Appelbaum, "Why Are Politicians So Obsessed With Manufacturing?." See also Shulevitz's argument for the Universal Basic Income in "It's Payback Time for Women."
122 Weber, *The Protestant Ethic and the "Spirit" of Capitalism.*
123 For more on anxiety related to drugs and pleasure, see Race, *Pleasure Consuming Medicine.* For more on anxiety related to dancing, see Craig, *Sorry I Don't Dance.* For a canonical treatment of moral panic, see Cohen, *Folk Devils and Moral Panics.*
124 For examples of scholarship on fandom and participatory culture, see Jenkins, *Fans, Bloggers, and Gamers*; Jenkins, *Convergence Culture*; Jenkins, *Textual Poachers.* For further examples of scholarship on the pro-social uses of social media, see Benkler, *The Wealth of Networks*; Shirky, *Here Comes Everybody*; Alexander Cho, "Queer Reverb: Tumblr, Affect, Time"; Castells, *Networks of Outrage and Hope.*
125 Gitlin, *Media Unlimited.*
126 Postman, *Amusing Ourselves to Death.*
127 Scott-Heron, *Small Talk at 125th and Lenox.*
128 Adorno, "A Social Critique of Radio Music."
129 Adorno, "On the Fetish Character in Music," 303, 307.
130 Ibid., 307.
131 For examples of utopianism, see Benkler, *The Wealth of Networks*; Jenkins, *Convergence Culture*; Papacharissi, "The Virtual Sphere." For accounts of early utopianism and dystopianism in writing about the Internet, see Sterne, "The Historiography of Cyberculture"; Wellman, "The Three Ages of Internet Studies."
132 For example, Evgeny Morozov, Sherry Turkle, and Jaron Lanier have all made this shift.
133 For a more in-depth analysis of these texts (and an earlier iteration of the present argument), see Goldberg, "Antisocial Media."
134 Safran Foer, "How Not to Be Alone."
135 Franzen, "Liking Is for Cowards"; Safran Foer, "How Not to Be Alone."
136 Scholz, *Uberworked and Underpaid*, 121.
137 Dean, *Blog Theory*, 77.
138 Ibid.
139 Ibid., 79.
140 Ibid., 125.

141 Ibid., 80.
142 Lovink, "What Is the Social in Social Media?"
143 Ibid.
144 See, for example, Adorno and Simpson, "On Popular Music"; Hall, "Encoding and Decoding in the Television Discourse."
145 See, for example, Katz, Blumler, and Gurevitch, "Uses and Gratifications Research."
146 One exception to this has been (some) scholarship on music and sound, perhaps a reflection of music as a boundary-blurring practice. See, for example, Daniel, "All Sound Is Queer."
147 Bersani, "Is the Rectum a Grave?," 212.
148 Ibid.
149 My intention here is not to reduce the varied meanings of "bottoming" to what Tan Hoang Nguyen labels, in *A View from the Bottom*, the "dominant perception," but rather to suggest that bottoming has been reviled according to a similar logic as being a passive media consumer. See also Scott, "Notes on Black (Power) Bottoms" in *Extravagant Abjection*.
150 Bersani, *Homos*, 76.
151 Morozov, *The Net Delusion*, 75. See also Scott, "Notes on Black (Power) Bottoms" in *Extravagant Abjection*.
152 Bolin, "Symbolic Production and Value in Media Industries."
153 Weeks, *The Problem with Work*, 162.

CHAPTER 3. AUTOMATING

1 Radliffe and Gavrilovic, "Are Robots Hurting Job Growth?"
2 Keynes, "Economic Possibilities for Our Grandchildren."
3 Brynjolfsson and McAfee, *The Second Machine Age*, 145.
4 Ibid., 146.
5 Ford, *The Lights in the Tunnel*, 162, 205.
6 See Aronowitz and Cutler, "Quitting Time: An Introduction," in *Post-Work*. Of course, immigration has also been a source of anxiety in terms of "job loss." In this arena too, discursive inconsistencies—immigrants are sometimes characterized both as lazy and as "stealing jobs"—suggest that interpretation is required.
7 Cutler, *Labor's Time*.
8 Brynjolfsson and McAfee, *The Second Machine Age*, 157.
9 Ford, *The Lights in the Tunnel*, 104.
10 Ibid., 155.
11 Ibid., 114.
12 See, for example, Wolf, "Enslave the Robots and Free the Poor"; "The Onrushing Wave."
13 See, for example, Brynjolfsson and McAfee, *The Second Machine Age*; Rampell, "Raging (Again) against the Robots."
14 Brynjolfsson and McAfee, *The Second Machine Age*, 216.

15  Ibid., 134.
16  Timberg, "Jaron Lanier: The Internet Destroyed the Middle Class."
17  Lanier, *Who Owns the Future?*, 2.
18  Payne, "Dear Marc Andreessen."
19  Ibid.
20  Parramore, "Don't Buy the Hype of a Robot-Driven 'Jobocalypse.'"
21  Ford, *The Lights in the Tunnel*, 167.
22  Ibid., 5.
23  Brynjolfsson and McAfee, *The Second Machine Age*, 232.
24  Ford, *The Lights in the Tunnel*, 173.
25  For a journalistic iteration of this argument, see Thompson, "A World without Work."
26  Payne, "Dear Marc Andreessen."
27  Ibid.
28  Timberg, "Jaron Lanier."
29  Ibid.
30  Ibid.
31  Lanier, "Sell Your Data to Save the Economy and Your Future."
32  Timberg, "Jaron Lanier."
33  Payne, "Dear Marc Andreessen."
34  Timberg, "Jaron Lanier."
35  Keynes, "Economic Possibilities for Our Grandchildren."
36  See Rampell, "Raging (Again) against the Robots."
37  See this book's Epilogue for a more in-depth exploration of this romanticization.
38  Parramore, "Don't Buy the Hype of a Robot-Driven 'Jobocalypse.'"
39  Ibid.
40  Solman, "Man vs. Machine."
41  Smith and Anderson, "AI, Robotics, and the Future of Jobs."
42  Rampell, "Raging (Again) against the Robots."
43  Ibid.
44  Condon and Wiseman, "Recession, Tech Kill Middle-Class Jobs."
45  Parramore, "Don't Buy the Hype of a Robot-Driven 'Jobocalypse.'" One might also consider here the general anxiety that surrounds technological development beyond automation. This might include, for example, the concern that humans are becoming robots, cyborgs, or post-human, and that the human is generally threatened by an encroachment of technology into private and civic life. See Jones, *Against Technology*.
46  Manjoo, "Will Robots Steal Your Job?"
47  Lohr, "Economists See More Jobs for Machines, Not People."
48  Wiseman and Condon, "Will Smart Machines Create a World without Work?"
49  Brynjolfsson and McAfee, *The Second Machine Age*, 91.
50  Wiseman and Condon, "Will Smart Machines Create a World without Work?"; Brynjolfsson and McAfee, *The Second Machine Age*, 29.

51  Ford, *The Lights in the Tunnel*, 99.
52  Brynjolfsson and McAfee, *The Second Machine Age*, 191.
53  "The Onrushing Wave."
54  Thompson, "What Jobs Will the Robots Take?"
55  Frey and Osborne, "The Future of Employment," 42.
56  Meltzer, "Robot Doctors, Online Lawyers and Automated Architects."
57  Ibid.
58  Manjoo, "Will Robots Steal Your Job?"
59  Ibid.
60  Miller, "Matt Miller: The Robots Are Coming."
61  Timberg, "Jaron Lanier."
62  Meltzer, "Robot Doctors, Online Lawyers and Automated Architects."
63  Ibid.
64  Wiseman and Condon, "Will Smart Machines Create a World without Work?"
65  See Brynjolfsson and McAfee, *The Second Machine Age*, 28–29.
66  Foucault, *The Order of Things*, 208.
67  Ibid., 213.
68  Ibid., 293.
69  Caffentzis, "Why Machines Cannot Create Value."
70  Ibid., 42.
71  Ibid., 53.
72  Ibid., 54.
73  Vahamaki and Virtanen, "Deleuze, Change, History."
74  Marx, *Capital, Volume One*, 186.
75  It is for this reason that Scholz insists on retaining the notion of work/labor in his analysis of what might otherwise be described as play or leisure. As he asks, "What would be lost when we give up on terms like labor and work? We would lose the legacy of the Triangle Shirtwaist Factory, Karen Silkwood, the strikes in Lordstown, Lawrence, and the Haymarket riots" (*Uberworked and Underpaid*, 107).
76  "Minimum Wage: Disappearing Jobs."
77  Ibid.
78  Manjoo, "Will Robots Steal Your Job?"
79  "The Onrushing Wave."
80  Ahmed, *Willful Subjects*, 15.
81  Ibid., 3.
82  Ibid., 146.
83  Ibid., 161.
84  Ibid., 7.
85  Ibid., 134.
86  Ibid., 12.
87  Ibid., 38.
88  Ibid., 174.

89 For many critics of automation, it is not enough for the human to be defined as willful in contrast to docile machines; the human must also be ethical. In other words, critics value the capacity of a human subject to resist in ethical ways. For an example of this discursive linking, see Carr, *The Glass Cage*, chap. 8.

90 Ahmed, *Willful Subjects*, 164.

91 Ibid., 160.

92 Ibid., 157.

93 Ibid., 169.

94 Ibid., 194.

95 For an example of this turn from refusal to collective governance, see Scholz's argument for "platform cooperativism" in *Uberworked and Underpaid*.

96 Consider, for example, Nicholas Carr's assertion that "machines are cold and mindless, and in their obedience to scripted routines we see an image of society's darker possibilities" (*The Glass Cage*, 21). Carr's use of "mindless" here indexes the kinds of cognitive or intellectual labor that machines are increasingly performing, but also machines' absence of will (that is, their "obedience"). In other words, to have a mind is not simply to be capable of cognitive/intellectual functioning, but to have the capacity for disobedience. To make matters worse, our increasing reliance on computers can cause "automation complacency," a cognitive ailment, according to Carr, which "takes hold when a computer lulls us into a false sense of security," sometimes with disastrous consequences (ibid., 67). So not only are machines docile, they are also impairing the human cognitive/intellectual faculties they are displacing. The word "complacency" here suggests an abdication of responsibility, again discursively connected to will or agency.

97 For more on the meanings of blue-collar labor, particularly in the context of the Great Recession, see the Epilogue.

98 Smith and Anderson, "AI, Robotics, and the Future of Jobs."

99 Brynjolfsson and McAfee, *The Second Machine Age*, 165.

100 Ibid., 120.

101 Marx and Engels, *The German Ideology*, 54.

102 "Chinese Factory Workers Fear They May Never Be Replaced with Machines."

103 Brynjolfsson and McAfee, *The Second Machine Age*, 256.

104 Smith and Anderson, "AI, Robotics, and the Future of Jobs."

105 Lanier, *Who Owns the Future?*, 9.

106 Brynjolfsson and McAfee, *The Second Machine Age*, 234.

107 Ibid.

108 Ibid.

109 Ibid., 237.

110 Ford, *The Lights in the Tunnel*, 169.

111 Ibid., 170.

112 Ibid., 194.

113 For another argument that suggests a link between work and happiness, see Carr, *The Glass Cage*. For an argument critical of this relation, see Frase, *Four Futures*.
114 Ford, *The Lights in the Tunnel*, 168.
115 Brynjolfsson and McAfee, *The Second Machine Age*, 257. This sentiment is echoed by psychologist Peter Gray, who argues that through play we "learn how to strategize, create new mental connections, express [our] creativity, cooperate, overcome narcissism, and get along with other people" (quoted in E. Strauss, "Would a Work-Free World Be So Bad?"). Here play is legitimated by its "harmony-promoting properties"; Gray imagines that a world with less work would be one in which more people participate in collective governance and generally display more pro-social behavior. The author of the *Atlantic* article in which Gray is interviewed (ibid.) also suggests that our currently impoverished forms of leisure—"beer and TV to a lot of Americans"—are a function of the displacement of older and presumably superior forms of play as a result of industrialization/urbanization. In "A World without Work" (also published in the *Atlantic*), Derek Thompson cites a number of scholars who similarly suggest that leisure will become more pro-social when work diminishes.
116 Brynjolfsson and McAfee, *The Second Machine Age*, 257.
117 Ibid.
118 Keynes, "Economic Possibilities for Our Grandchildren," 326.
119 Ibid.
120 Ibid., 327.
121 Ibid., 328.
122 Ibid., 327, 328.
123 Ibid., 329.
124 Ibid., 330.
125 Ibid., 327.
126 As previously alluded to, Frase's communist vision of a post-work future (which cites Keynes's own vision) is also crafted in this mold. See Frase, *Four Futures*.
127 Piven and Cloward, *Regulating the Poor*, 6.
128 As quoted in Kreider, "The 'Busy' Trap."
129 Ibid.

CHAPTER 4. SHARING
1 "How TaskRabbit Gave Jennifer G. Her Life Back."
2 Botsman and Rogers, *What's Mine Is Yours*, xv–xvi.
3 Gansky, *The Mesh*.
4 "Peers Blog"; Kamenetz, "Is Peers the Sharing Economy's Future or Just a Great Silicon Valley PR Stunt?"
5 "Peers: Growing the Sharing Economy Movement."
6 Scholz, *Uberworked and Underpaid*, 55.
7 Horning, "'Sharing' Economy and Self-Exploitation."
8 Singer, "In the Sharing Economy, Workers Find Both Freedom and Uncertainty."

9   See Federici, "Precarious Labor."

10  Kessler, "Pixel and Dimed."

11  Ibid.

12  Singer, "In the Sharing Economy."

13  Ibid.

14  Leonard, "Why Uber Must Be Stopped."

15  Ibid. Leonard's prediction was prescient as Uber began testing driverless cars in 2016. See Kang, "No Driver?"

16  Morozov, "The 'Sharing Economy' Undermines Workers' Rights."

17  Singer, "In the Sharing Economy."

18  Morozov, "The 'Sharing Economy' Undermines Workers' Rights."

19  Asher-Schapiro, "Against Sharing."

20  Cagle, "The Case against Sharing."

21  Singer, "In the Sharing Economy."

22  Ibid.

23  Morozov, "The 'Sharing Economy' Undermines Workers' Rights."

24  Cagle, "The Case against Sharing."

25  Friedman, "Welcome to the 'Sharing Economy'"; Kessler, "Pixel and Dimed."

26  Morozov, "Don't Believe the Hype, the 'Sharing Economy' Masks a Failing Economy."

27  Roose, "The Sharing Economy Isn't about Trust, It's about Desperation."

28  Ibid.

29  For more on sharing in the context of early 2000s file-sharing, see Goldberg, "Own Nothing, Have Everything."

30  Scholz, "The Politics of the Sharing Economy."

31  For an analysis of sharing economy rhetoric that challenges both proponents' use of "sharing" and critics' challenges to this use, see John, *The Age of Sharing*, chap. 4.

32  Singer, "In the Sharing Economy."

33  Scheiber, "Silicon Valley Is Ruining 'Sharing' for Everybody."

34  Ibid.

35  Scholz, "The Politics of the Sharing Economy."

36  Asher-Schapiro, "Against Sharing."

37  Finch, "Renting Out Your Property Is Not 'Sharing.'"

38  Ibid.

39  Leonard, "Millennials Will Not Be Regulated."

40  Ibid.

41  Horning, "'Sharing' Economy and Self-Exploitation."

42  North, "The Dark Side of Sharing."

43  Fung, "Yes, Airbnb's New Logo Looks like a Butt. That's Kind of the Point."

44  North, "The Dark Side of Sharing."

45  That said, proponents of the sharing economy tend to be far less critical of the market and of acting in one's self-interest. For example, in *The Mesh* Gansky

writes, "As an entrepreneur, I'm excited to have a new platform to reinvent markets and create thriving, customer-loving businesses. But I'm also thankful for new approaches that are good for the planet and its inhabitants."

46  Horning, "'Sharing' Economy and Self-Exploitation."

47  Orsi, "The Sharing Economy Just Got Real."

48  Scholz, "The Politics of the Sharing Economy." Elsewhere Scholz similarly asks, "How can we talk about genuine sharing or innovation when a third party immediately monetizes your every interaction for the benefit of a small group of stockholders?" and then laments, "What used to be favors among friends now has a price tag. . . . Shares are shared and sharing becomes shearing" (*Uberworked and Underpaid*, 43, 46).

49  Scholz, "The Politics of the Sharing Economy."

50  Ibid.

51  Ibid.

52  Hesmondhalgh, "User-Generated Content, Free Labour and the Cultural Industries," 276.

53  Ibid., 274.

54  Ibid., 278.

55  Ibid., 277.

56  Ibid., 278.

57  Ibid.

58  As Scholz writes, following Kathi Weeks, "The refusal of work is really a refusal of the way work is organized" (*Uberworked and Underpaid*, 153).

59  As quoted in Hyde, *The Gift*, 87.

60  As Scholz writes, "Free labor itself is not a problem outside of extractive platform capitalism; it can in fact be a site of resistant subjectivities and emerging forms of solidarity" (*Uberworked and Underpaid*, 76).

61  Fuchs, "Class and Exploitation on the Internet," 221.

62  Terranova, "Free Labor," 48.

63  Michel Bauwens, "Thesis on Digital Labor in an Emerging P2P Economy," 209.

64  This dynamic is satirized in an episode of the television show *Black Mirror* ("Nosedive"), in which people continually rank others on a scale of one to five. High scores are required to access various housing, employment, and networking opportunities.

65  Scholz, "Introduction: Why Does Digital Labor Matter Now?," 2; Terranova, "Free Labor," 44.

66  Marx and Engels, *The Communist Manifesto*, 223.

67  Ibid., 222.

68  For a canonical treatment of this Marx and Engels passage see Berman, *All That Is Solid Melts into Air*. Berman was particularly interested in the disintegration described in the passage, as well as in the dialectical dynamic by which social and political-economic forms—for Berman, in modernity in particular—contain the seeds of their own destruction.

69 Ibid.
70 Ibid.
71 Ibid., 223.
72 Tronti, "Workers and Capital."
73 Simmel, *The Philosophy of Money*, 445.
74 Ibid., 310.
75 Ibid., 213.
76 Taylor, *Confidence Games*, 59.
77 Simmel, *The Philosophy of Money*, 298.
78 Ibid., 257.
79 Zelizer, *The Social Meaning of Money*, 6.
80 Ibid., 11.
81 Ibid., 18.
82 Ibid., 30.
83 Ibid., 64.
84 Ibid., 90–91.
85 Ibid., 214.
86 Simmel, *The Philosophy of Money*, 298.
87 Ibid., 246.
88 Ibid.
89 Ibid., 344.
90 Ibid., 337.
91 Ibid., 344.
92 Ibid.
93 Zelizer, *The Social Meaning of Money*, 52.
94 D'Emilio, "Capitalism and Gay Identity."
95 Hyde, *The Gift*, 69.
96 Ibid., 58.
97 Simmel, *The Philosophy of Money*, 345.
98 Ibid., 225.
99 Ibid., 482.
100 Ibid., 226.
101 Ibid., 445.
102 Hyde, *The Gift*, 81.
103 Taylor, *Confidence Games*, 60.
104 Ibid., 73–74.
105 Hyde, *The Gift*, 70.
106 Ibid., 75.
107 Frase, *Four Futures*, 47.
108 Orsi, "The Sharing Economy Just Got Real." For a detailed description of and argument for what Scholz calls "platform cooperativism," see *Uberworked and Underpaid*, chap. 7.
109 Ibid., 90.

110 Hyde, *The Gift*, 83.
111 Horning, "'Sharing' Economy and Self-Exploitation."
112 Ibid.
113 Ibid.
114 Hyde, *The Gift*, 47.
115 Ibid., 53.
116 Ibid., 48.
117 Ibid., 52.
118 Ibid., 51.
119 For more on the conceptual differentiation of work from labor, see Scholz, *Uberworked and Underpaid*, chap. 3.
120 Hyde, *The Gift*, 28.
121 Ibid., 91.
122 Ibid., 9.
123 Ibid., 88.
124 As Nicholas Carr laments, "The most meaningful bonds aren't forged through transaction in a marketplace or other routinized exchanges. . . . The bonds require trust and courtesy and sacrifice, all of which, at least to a technocrat's mind, are sources of inefficiency and inconvenience. Removing the friction from social attachments doesn't strengthen them; it weakens them. It makes them more like the attachments between consumers and products—easily formed and just as easily broken" (*The Glass Cage*, 181). The rather ambiguous word "meaningful" here serves as an evaluative standard of relations (or "bonds") which lack trust, courtesy, and sacrifice and are apparently less meaningful and therefore less worthy.
125 Federici, "Precarious Labor."
126 If this point has been lost on critics, it seems to have been noted by sharing economy companies. For example, a *New Yorker* article quotes Talmon Marco (CEO of Juno, a rival of Uber), who describes how an independent-contractor model allows Juno to "take joint custody" over workers who would otherwise be unavailable. The primary way Juno does this is by offering workers higher commissions and better perks. See Kolhatkar, "The Anti-Uber."
127 Singer, "In the Sharing Economy."
128 Scheiber, "Silicon Valley Is Ruining 'Sharing' for Everybody."
129 Josh MacPhee, "Who's the Shop Steward on Your Kickstarter?"
130 Ibid.
131 Kelly, "People Want to Pay."

EPILOGUE
1 Marx and Engels, *The Communist Manifesto*, 223.
2 Shakespeare, *The Tempest*, 53.
3 Ferguson, *Inside Job*.
4 Rich and Leonhardt, "Trading Places: Real Estate Instead of Dot-Coms."
5 Jonze, *Her*.

6 "Barack Obama's Inaugural Address."
7 Leonhardt, "After the Great Recession."
8 See Appelbaum, "Why Are Politicians So Obsessed With Manufacturing?"; Porter, "The Mirage of a Return to Manufacturing Greatness."
9 Binkley, "Designers Mine American Heritage for Rags and Riches."
10 Garbarino, "Is L. L. Bean Driving the Runway?!"
11 "Levi's Proclaims 'We Are All Workers.'"
12 See Appelbaum, "Why Are Politicians So Obsessed With Manufacturing?"
13 Klein, *No Logo: Taking Aim at the Brand Bullies*, 3–4.
14 Maslow, "A Theory of Human Motivation."
15 See Marcuse, *One-Dimensional Man*; Schoenberg, *Style and Idea.*
16 Derrida, *Specters of Marx*; Clough, *Autoaffection.*
17 Marx and Engels, *The Communist Manifesto*, 223.
18 Durkheim, *The Elementary Forms of the Religious Life.*
19 The insistence that we reclassify what seems to be immaterial as, in fact, material (or real) might thus be interpreted as an assertion of the importance of the social bonds discursively attached to the material, as when Scholz insists that "digital labor is everything but 'immaterial;' it is a sector of the economy, a set of human activities that is predicated on global supply chains of sweated material labor" (*Uberworked and Underpaid*, 99).

# BIBLIOGRAPHY

Adorno, Theodor W. "A Social Critique of Radio Music." *Kenyon Review* 7, no. 2 (1945): 208–17.

——. "On the Fetish Character in Music and the Regression of Listening." In *Essays on Music*, edited by Richard Leppert, translated by Susan H. Gillespie, 288–317. Berkeley: University of California Press, 2002.

Adorno, Theodor W., and George Simpson. "On Popular Music." In *Essays on Music*, edited by Richard Leppert, translated by Susan H. Gillespie, 437–69. Berkeley: University of California Press, 2002.

Ahmed, Sara. *Willful Subjects*. Durham, NC: Duke University Press, 2014.

——. *The Cultural Politics of Emotion*. New York: Routledge, 2012.

——. "Orientations: Toward a Queer Phenomenology." *GLQ: A Journal of Lesbian and Gay Studies* 12, no. 4 (2006): 543–74.

——. *Queer Phenomenology: Orientations, Objects, Others*. Durham, NC: Duke University Press, 2006.

——. "Collective Feelings, or, the Impressions Left by Others." *Theory, Culture & Society* 21, no. 2 (April 1, 2004): 25–42.

——. "Declarations of Whiteness: The Non-Performativity of Anti-Racism." *Borderlands* 3, no. 2 (2004).

Andrejevic, Mark. "Estranged Free Labor." In *Digital Labor: The Internet as Playground and Factory*, edited by Trebor Scholz, 149–64. New York: Routledge, 2013.

Andrejevic, Mark, John Banks, John Edward Campbell, Nick Couldry, Adam Fish, Alison Hearn, and Laurie Ouellette. "Participations: Dialogues on the Participatory Promise of Contemporary Culture and Politics." *International Journal of Communication* 8 (2014): 1089–1106.

Appelbaum, Binyamin. "Why Are Politicians So Obsessed with Manufacturing?" *New York Times*, October 4, 2016. www.nytimes.com.

Aronowitz, Stanley, and Jonathan Cutler. *Post-Work: The Wages of Cybernation*. New York: Routledge, 1998.

Arvidsson, Adam. "On the 'Pre-History of the Panoptic Sort': Mobility in Market Research." *Surveillance & Society* 1, no. 4 (2004): 456–74.

Asher-Schapiro, Avi. "Against Sharing." *Jacobin*, September 19, 2014. www.jacobinmag.com.

"Barack Obama's Inaugural Address." *New York Times*, January 20, 2009. www.nytimes. com.

Barad, Karen. *Meeting the Universe Halfway: Quantum Physics and the Entanglement of Matter and Meaning*. Durham, NC: Duke University Press, 2007.

Barlow, David H. *Anxiety and Its Disorders: The Nature and Treatment of Anxiety and Panic*. New York: Guilford Press, 2004.

Battelle, John, and Tim O'Reilly. "Web 2.0 Summit," n.d. www.web2summit.com.

Baudrillard, Jean. *Simulacra and Simulation*. Translated by Sheila Faria Glaser. Ann Arbor: University of Michigan Press, 1994.

Bauwens, Michel. "Thesis on Digital Labor in an Emerging P2P Economy." In *Digital Labor: The Internet as Playground and Factory*, edited by Trebor Scholz, 207–10. New York: Routledge, 2013.

Beller, Jonathan. "Capital/Cinema." In *Deleuze and Guattari: New Mappings in Politics, Philosophy, and Culture*, edited by Eleanor Kaufman and Kevin Jon Heller, 77–95. Minneapolis: University of Minnesota Press, 1998.

Benkler, Yochai. *The Wealth of Networks: How Social Production Transforms Markets and Freedom*. New Haven, CT: Yale University Press, 2006.

Berlant, Lauren Gail. *Cruel Optimism*. Durham, NC: Duke University Press, 2011.

Bersani, Leo. "Sociability and Cruising." *UMBR(a)*, "Sameness" issue, no. 1 (2002): 9–23.

———. *Homos*. Cambridge, MA: Harvard University Press, 1996.

———. "Is the Rectum a Grave?" *October* 43 (1987): 197–222.

Bersani, Leo, and Adam Phillips. *Intimacies*. Chicago: University of Chicago Press, 2008.

Best, Stephen, and Sharon Marcus. "Surface Reading: An Introduction." *Representations* 108, no. 1 (2009): 1–21.

Binkley, Christina. "Designers Mine American Heritage for Rags and Riches." *Wall Street Journal*, October 15, 2009. www.wsj.com.

Bogost, Ian. "Gamification Is Bullshit." *Ian Bogost*, August 8, 2011. www.bogost.com.

———. "Persuasive Games: Exploitationware." *Gamasutra*, May 3, 2011. www.gamasutra.com.

Bolin, Göran. "Symbolic Production and Value in Media Industries." *Journal of Cultural Economy* 2, no. 3 (2009): 345–61.

Botsman, Rachel, and Roo Rogers. *What's Mine Is Yours: The Rise of Collaborative Consumption*. New York: HarperCollins, 2010.

Berman, Marshall. *All That Is Solid Melts into Air: The Experience of Modernity*. New York: Penguin, 1988.

Browne, Nick. "The Political Economy of the Television (Super) Text." *Quarterly Review of Film Studies* 9, no. 3 (1984): 174–82.

Bruns, Axel. *Blogs, Wikipedia, Second Life, and Beyond: From Production to Produsage*. New York: Peter Lang, 2008.

Brynjolfsson, Erik, and Andrew McAfee. *The Second Machine Age: Work, Progress, and Prosperity in a Time of Brilliant Technologies*. New York: W. W. Norton, 2014.

Butler, Judith. "Remarks on 'Queer Bonds.'" *GLQ: A Journal of Lesbian and Gay Studies* 17, nos. 2–3 (2011): 381–87.

Caffentzis, C. George. "Why Machines Cannot Create Value; Or, Marx's Theory of Machines." In *Cutting Edge: Technology, Information Capitalism and Social Revolution*, edited by Jim Davis, Thomas Hirschl, and Michael Stack, 29–56. New York: Verso, 1997.

Cagle, Susie. "The Case against Sharing." *Nib*, May 27, 2014. www.medium.com.

Carr, Nicholas. *The Glass Cage: Where Automation Is Taking Us*. New York: Random House, 2015.

———. *The Big Switch: Rewiring the World, from Edison to Google*. New York: W. W. Norton, 2008.

Caserio, Robert L., Lee Edelman, Judith Halberstam, José Esteban Muñoz, and Tim Dean. "The Antisocial Thesis in Queer Theory." *PMLA* 121, no. 3 (2006): 819–28.

Castells, Manuel. *Networks of Outrage and Hope: Social Movements in the Internet Age*. Malden, MA: Polity Press, 2012.

"Chinese Factory Workers Fear They May Never Be Replaced with Machines." *Onion*, March 19, 2014. www.theonion.com.

Cho, Alexander. "Queer Reverb: Tumblr, Affect, Time." In *Networked Affect*, edited by Ken Hills, Susanna Paasonen, and Michael Petit, 43–58. Cambridge, MA: MIT Press, 2015.

Chow, Rey. *Writing Diaspora: Tactics of Intervention in Contemporary Cultural Studies*. Bloomington: Indiana University Press, 1993.

Clough, Patricia Ticineto. *Autoaffection: Unconscious Thought in the Age of Teletechnology*. Minneapolis: University of Minnesota Press, 2000.

Clough, Patricia Ticineto, and Jean Halley, eds. *The Affective Turn: Theorizing the Social*. Durham, NC: Duke University Press, 2007.

Cohen, Stanley. *Folk Devils and Moral Panics: The Creation of the Mods and Rockers*. London: MacGibbon and Kee, 1972.

Condon, Bernard, and Paul Wiseman. "Recession, Tech Kill Middle-Class Jobs." Associated Press, January 25, 2013. www.bigstory.ap.org.

Craig, Maxine Leeds. *Sorry I Don't Dance: Why Men Refuse to Move*. New York: Oxford University Press, 2013.

Crapanzano, Vincent. *Hermes' Dilemma and Hamlet's Desire: On the Epistemology of Interpretation*. Cambridge, MA: Harvard University Press, 1992.

Crary, Jonathan. *24/7: Late Capitalism and the Ends of Sleep*. New York: Verso, 2013.

Cutler, Jonathan. *Labor's Time: Shorter Hours, the UAW, and the Struggle for American Unionism*. Philadelphia: Temple University Press, 2008.

Cvetkovich, Ann. *Depression: A Public Feeling*. Durham, NC: Duke University Press, 2012.

Daniel, Drew. "All Sound Is Queer." *Wire* 33, no. 3 (2011): 43–6.

De Lauretis, Teresa. "Queer Texts, Bad Habits, and the Issue of a Future." *GLQ: A Journal of Lesbian and Gay Studies* 17, nos. 2–3 (2011): 243–63.

Dean, Jodi. *Blog Theory: Feedback and Capture in the Circuits of Drive*. Malden, MA: Polity Press, 2010.

Dean, Tim. "No Sex Please, We're American." *American Literary History* 27, no. 3 (2015): 614–24.

———. *Unlimited Intimacy: Reflections on the Subculture of Barebacking*. Chicago: University of Chicago Press, 2009.

D'Emilio, John. "Capitalism and Gay Identity." In *The Gender/Sexuality Reader: Culture, History, Political Economy*, edited by Roger N. Lancaster and Micaela di Leonardo, 169–78. New York: Routledge, 1997.

Derrida, Jacques. *Specters of Marx: The State of the Debt, the Work of Mourning and the New International*. Translated by Peggy Kamuf. New York: Routledge, 1994.

Dienst, Richard. *Still Life in Real Time: Theory after Television*. Durham, NC: Duke University Press, 1994.

"The Dot-Com Bubble Bursts." *New York Times*, December 24, 2000. www.nytimes.com.

Duggan, Lisa. "Queer Complacency without Empire." *Bully Bloggers*, September 22, 2015. www.bullybloggers.wordpress.com.

———. "The New Homonormativity: The Sexual Politics of Neoliberalism." In *Materializing Democracy: Toward a Revitalized Cultural Politics*, edited by Russ Castronovo and Dana D. Nelson, 175–94. Durham, NC: Duke University Press, 2002.

Durkheim, Émile. *The Elementary Forms of the Religious Life*. Translated by Joseph Ward Swain. London: George Allen and Unwin, 1915.

Dyer-Witheford, Nick. "Digital Labour, Species-Becoming and the Global Worker." *Ephemera: Theory & Politics in Organization* 10, no. 3 (2010): 484–503.

Dyer-Witheford, Nick, and Greig De Peuter. *Games of Empire: Global Capitalism and Video Games*. Minneapolis: University of Minnesota Press, 2009.

Edelman, Lee. *No Future: Queer Theory and the Death Drive*. Durham, NC: Duke University Press, 2004.

Eng, David L., Judith Halberstam, and José Esteban Muñoz. "Introduction: What's Queer about Queer Studies Now?" *Social Text* 23, no. 3-4 84–85 (2005): 1–17.

Federici, Silvia. "Precarious Labor: A Feminist Viewpoint." *In the Middle of a Whirlwind: 2008 Convention Protests, Movement and Movements*, 2008. www.joaap.org.

Ferguson, Charles. *Inside Job*. Culver City, CA: Sony Pictures Classics, 2010.

Ferguson, Roderick A. *Aberrations in Black: Toward a Queer of Color Critique*. Minneapolis: University of Minnesota Press, 2004.

Finch, Kevin. "Renting Out Your Property Is Not 'Sharing.'" *KevinPDX*, August 15, 2014. www.kevinpdx.com.

Fitzpatrick, Alex. "This One Stat Reveals the Sharing Economy's Racism Problem." *Time*, December 14, 2015. www.time.com.

Ford, Martin. *The Lights in the Tunnel: Automation, Accelerating Technology and the Economy of the Future*. N.p.: Acculant Publishing, 2009.

Foucault, Michel. *The Order of Things: An Archaeology of the Human Sciences*. London: Tavistock, 1970.

Franzen, Jonathan. "Liking Is for Cowards. Go for What Hurts." *New York Times*, May 28, 2011. www.nytimes.com.

Frase, Peter. *Four Futures: Life after Capitalism*. New York: Verso, 2016.

Frey, Carl Benedikt, and Michael A. Osborne. "The Future of Employment: How Susceptible Are Jobs to Computerisation?" Oxford: Oxford Martin School, September 2013. www.oxfordmartin.ox.ac.uk.

Friedman, Thomas L. "Welcome to the 'Sharing Economy.'" *New York Times*, July 21, 2013. www.nytimes.com.

Fuchs, Christian. "Class and Exploitation on the Internet." In *Digital Labor: The Internet as Playground and Factory*, edited by Trebor Scholz, 211–24. New York: Routledge, 2012.

Fung, Brian. "Yes, Airbnb's New Logo Looks like a Butt. That's Kind of the Point." *Washington Post*, July 16, 2014. www.washingtonpost.com.

Gansky, Lisa. *The Mesh: Why the Future of Business Is Sharing*. New York: Penguin, 2010.

Garbarino, Steve. "Is L. L. Bean Driving the Runway?!" *Wall Street Journal*, September 25, 2010. www.wsj.com.

Gitlin, Todd. *Media Unlimited: How the Torrent of Images and Sounds Overwhelms Our Lives*. New York: Holt, 2007.

Glassner, Barry. "The Construction of Fear." *Qualitative Sociology* 22, no. 4 (1999): 301–9.

Goldberg, Greg. "Antisocial Media: Digital Dystopianism as a Normative Project." *New Media & Society* 18, no. 5 (2016): 784–99.

———. "Own Nothing, Have Everything: Peer-to-Peer Networks and the New Cultural Economy." PhD diss., City University of New York, 2009.

Gregg, Melissa, and Gregory J. Seigworth, eds. *The Affect Theory Reader*. Durham, NC: Duke University Press, 2010.

Guha, Ranajit. *Elementary Aspects of Peasant Insurgency in Colonial India*. Durham, NC: Duke University Press, 1999.

Halberstam, Jack. "Straight Eye for the Queer Theorist—A Review of 'Queer Theory without Antinormativity.'" *Bully Bloggers*, September 12, 2015. www.bullybloggers.wordpress.com.

Halberstam, Judith. *The Queer Art of Failure*. Durham, NC: Duke University Press, 2011.

———. "Shame and White Gay Masculinity." *Social Text* 23 (Fall–Winter 2005): 219–33.

Hall, Stuart. "Encoding and Decoding in the Television Discourse." Birmingham: Centre for Cultural Studies, University of Birmingham, 1973.

Halperin, David M., and Valerie Traub, eds. *Gay Shame*. Chicago: University of Chicago Press, 2009.

Hardt, Michael, and Antonio Negri. *Multitude: War and Democracy in the Age of Empire*. New York: Penguin, 2004.

Hesmondhalgh, David. "User-Generated Content, Free Labour and the Cultural Industries." *Ephemera* 10, no. 3/4 (2010): 267–84.

Hochschild, Arlie, and Anne Machung. *The Second Shift: Working Families and the Revolution at Home*. New York: Viking Penguin, 1989.

hooks, bell. "Eating the Other: Desire and Resistance." In *Black Looks: Race and Representation*, 21–39. New York: South End Press, 1992.

Horning, Rob. "'Sharing' Economy and Self-Exploitation." *New Inquiry*, June 20, 2014. www.thenewinquiry.com.

"How TaskRabbit Gave Jennifer G. Her Life Back," 2012. www.youtube.com.

Hyde, Lewis. *The Gift: Creativity and the Artist in the Modern World*. Edinburgh: Canongate Books, 2007.

Jenkins, Henry. *Textual Poachers: Television Fans and Participatory Culture*. New York: Routledge, 2012.

———. *Convergence Culture: Where Old and New Media Collide*. New York: New York University Press, 2006.

———. *Fans, Bloggers, and Gamers: Exploring Participatory Culture*. New York: New York University Press, 2006.

Jhally, Sut. *The Factory in the Living Room*, 2012. www.vimeo.com.

John, Nicholas A. *The Age of Sharing*. Malden, MA: Polity Press, 2017.

Jones, Steven E. *Against Technology: From the Luddites to Neo-Luddism*. New York: Routledge, 2013.

Jonze, Spike. *Her*. Burbank, CA: Warner Bros., 2013.

Joseph, Miranda. *Against the Romance of Community*. Minneapolis: University of Minnesota Press, 2002.

Kamenetz, Anya. "Is Peers the Sharing Economy's Future or Just a Great Silicon Valley PR Stunt?" *Fast Company*, December 9, 2013. www.fastcompany.com.

Kang, Cecilia. "No Driver? Bring It On. How Pittsburgh Became Uber's Testing Ground." *New York Times*, September 10, 2016. www.nytimes.com.

Katz, Elihu, Jay G. Blumler, and Michael Gurevitch. "Uses and Gratifications Research." *Public Opinion Quarterly* 37, no. 4 (1973): 509–23.

Kelly, Kevin. "People Want to Pay." *Technium*, August 1, 2008. www.kk.org.

Kessler, Sarah. "Pixel and Dimed: On (Not) Getting By in the Gig Economy." *Fast Company*, May 2014. www.fastcompany.com.

Keynes, John Maynard. "Economic Possibilities for Our Grandchildren." In *Essays in Persuasion*, 358–74. New York: Palgrave Macmillan, 2010.

Kierkegaard, Søren. *The Concept of Dread*. Translated by Walter Lowrie. Princeton, NJ: Princeton University Press, 1957.

Klein, Naomi. *No Logo: Taking Aim at the Brand Bullies*. New York: Picador, 2000.

Kolhatkar, Sheelah. "The Anti-Uber." *New Yorker*, October 10, 2016.

Kreider, Tim. "The 'Busy' Trap." *New York Times*, June 30, 2012. www.opinionator. blogs.nytimes.com.

Kücklich, Julian. "Precarious Playbour: Modders and the Digital Games Industry." *The Fibreculture Journal* 5 (2005).

Lanier, Jaron. *Who Owns the Future?* New York: Simon and Schuster, 2014.

———. "Sell Your Data to Save the Economy and Your Future." *BBC News*, May 23, 2013. www.bbc.co.uk.

Leonard, Andrew. "Why Uber Must Be Stopped." *Salon*, August 31, 2014. www.salon. com.

———. "Millennials Will Not Be Regulated." *Salon*, September 20, 2013. www.salon. com.

Leonhardt, David. "After the Great Recession." *New York Times*, April 28, 2009. www. nytimes.com.

"Levi's Proclaims 'We Are All Workers' with Launch of Latest Go Forth Marketing Campaign." *PR NewsWire*, June 24, 2010. www.prnewswire.com.

Lohr, Steve. "Economists See More Jobs for Machines, Not People." *New York Times*, October 23, 2011, New York edition. www.nytimes.com.

Love, Heather. *Feeling Backward: Loss and the Politics of Queer History*. Cambridge, MA: Harvard University Press, 2009.

Lovink, Geert. "What Is the Social in Social Media?" *E-Flux* 40 (December 2012). www.e-flux.com.

MacPhee, Josh. "Who's the Shop Steward on Your Kickstarter?" *Baffler*, 2012. www. thebaffler.com.

Manjoo, Farhad. "Will Robots Steal Your Job?" *Slate*, September 26, 2011. www.slate. com.

Marcuse, Herbert. *One-Dimensional Man: Studies in the Ideology of Advanced Industrial Society*. Boston: Beacon Press, 2012.

Marino, Gordon D. "Anxiety in 'The Concept of Anxiety.'" In *The Cambridge Companion to Kierkegaard*, edited by Alastair Hannay and Gordon D. Marino, 308–28. New York: Cambridge University Press, 1998.

Marx, Karl. *Capital, Volume One: A Critique of Political Economy*. Edited by Friedrich Engels. Translated by Samuel Moore and Edward Aveling. Mineola, NY: Dover, 2011.

Marx, Karl, and Friedrich Engels. *The Communist Manifesto*. Translated by Samuel Moore. New York: Penguin, 2002.

———. *The German Ideology*. Edited by Christopher John Arthur. New York: International Publishers, 1970.

Maslow, Abraham H. "A Theory of Human Motivation." *Psychological Review* 50, no. 4 (1943): 370–96.

Meltzer, Tom. "Robot Doctors, Online Lawyers and Automated Architects: The Future of the Professions?" *Guardian*, June 15, 2014. www.theguardian.com.

Menon, Madhavi. *Indifference to Difference: On Queer Universalism*. Minneapolis: University Of Minnesota Press, 2015.

———. "Universalism and Partition: A Queer Theory." *Differences* 26, no. 1 (2015): 117–40.

Merleau-Ponty, Maurice. *Phenomenology of Perception*. Translated by Colin Smith. New York: Routledge, 2005.

Miller, Matt. "Matt Miller: The Robots Are Coming." *Washington Post*, January 9, 2013. www.washingtonpost.com.

"Minimum Wage: Disappearing Jobs," 2014. www.youtube.com.

Monroe, Rachel. "More Guests, Empty Houses." *Slate*, February 13, 2014. www.slate. com.

Morozov, Evgeny. "Don't Believe the Hype, the 'Sharing Economy' Masks a Failing Economy." *Guardian*, September 27, 2014. www.theguardian.com.

———. "The 'Sharing Economy' Undermines Workers' Rights." *Financial Times*, October 14, 2013. www.ft.com.

———. *The Net Delusion: The Dark Side of Internet Freedom*. New York: PublicAffairs, 2011.

Muñoz, José Esteban. *Cruising Utopia: The Then and There of Queer Futurity*. New York: New York University Press, 2009.

Newcomer, Eric, and Olivia Zaleski. "Inside Uber's Auto-Lease Machine, Where Almost Anyone Can Get a Car." *Bloomberg Technology*, May 31, 2016. www.bloomberg.com.

Ngai, Sianne. *Ugly Feelings*. Cambridge, MA: Harvard University Press, 2005.

Nguyen, Tan Hoang. *A View from the Bottom: Asian American Masculinity and Sexual Representation*. Durham, NC: Duke University Press, 2014.

North, Anna. "The Dark Side of Sharing." *New York Times*, July 22, 2014. www.op-talk. blogs.nytimes.com.

"Obama's 2013 State of the Union Address." *New York Times*, February 12, 2013. www. nytimes.com.

"The Onrushing Wave." *Economist*, January 18, 2014. www.economist.com.

O'Reilly, Tim, and John Battelle. "Web Squared: Web 2.0 Five Years On," 2009. www. web2summit.com.

Orr, Jackie. *Panic Diaries: A Genealogy of Panic Disorder*. Durham, NC: Duke University Press, 2006.

Orsi, Janelle. "The Sharing Economy Just Got Real." Shareable, September 16, 2013. www.shareable.net.

Papacharissi, Zizi. "The Virtual Sphere: The Internet as a Public Sphere." *New Media & Society* 4, no. 1 (2002): 9–27.

Parramore, Lynn Stuart. "Don't Buy the Hype of a Robot-Driven 'Jobocalypse.'" *Al Jazeera America*, February 7, 2014. www.america.aljazeera.com.

Patterson, Orlando. *Slavery and Social Death*. Cambridge, MA: Harvard University Press, 1982.

Payne, Alex. "Dear Marc Andreessen." *Valleywag*, June 18, 2014. www.valleywag. gawker.com.

"Peers Blog." *Peers*, n.d. www.blog.peers.org.

"Peers: Growing the Sharing Economy Movement," 2013. www.youtube.com.

Peiss, Kathy. *Cheap Amusements: Working Women and Leisure in Turn-of-the-Century New York*. Philadelphia: Temple University Press, 1986.

Piven, Frances Fox, and Richard Cloward. *Regulating the Poor: The Functions of Public Welfare*. New York: Vintage, 1993.

"Points of Control: The Battle for the Internet Economy," 2010. www.youtube.com.

Porter, Eduardo. "The Mirage of a Return to Manufacturing Greatness." *New York Times*, April 26, 2016. www.nytimes.com.

Postigo, Hector. "From Pong to Planet Quake: Post-Industrial Transitions from Leisure to Work." *Information, Communication & Society* 6, no. 4 (2003): 593–607.

Postman, Neil. *Amusing Ourselves to Death: Public Discourse in the Age of Show Business*. New York: Penguin, 2006.

Race, Kane. *Pleasure Consuming Medicine: The Queer Politics of Drugs*. Durham, NC: Duke University Press, 2009.

Radliffe, Harry A., II, and Maria Gavrilovic. "Are Robots Hurting Job Growth?" *60 Minutes*. CBS, January 13, 2013. www.cbsnews.com.

Raffoul, François. *The Origins of Responsibility*. Bloomington: Indiana University Press, 2010.

Rampell, Catherine. "Raging (Again) against the Robots." *New York Times*, February 2, 2013. www.nytimes.com.

Rich, Motoko, and David Leonhardt. "Trading Places: Real Estate Instead of Dot-Coms." *New York Times*, March 25, 2005. www.nytimes.com.

Ritzer, George, and Nathan Jurgenson. "Production, Consumption, Prosumption: The Nature of Capitalism in the Age of the Digital 'Prosumer.'" *Journal of Consumer Culture* 10, no. 1 (March 1, 2010): 13–36.

Roemer, John E. *Egalitarian Perspectives: Essays in Philosophical Economics*. New York: Cambridge University Press, 1996.

Roose, Kevin. "The Sharing Economy Isn't about Trust, It's about Desperation." *New York Magazine*, April 24, 2014. www.nymag.com.

Ross, Andrew. "In Search of a Lost Paycheck." In *Digital Labor: The Internet as Playground and Factory*, edited by Trebor Scholz, 13–32. New York: Routledge, 2013.

———. "No-Collar Labour in America's 'New Economy.'" *Socialist Register* 37 (2009): 77–87.

———. *No-Collar: The Humane Workplace and Its Hidden Costs*. New York: Basic, 2003.

Safran Foer, Jonathan. "How Not to Be Alone." *New York Times*, June 8, 2013. www.nytimes.com.

Saunders, George. "Who Are All These Trump Supporters?" *New Yorker*, July 11, 2016. www.newyorker.com.

Scheiber, Noam. "Silicon Valley Is Ruining 'Sharing' for Everybody." *New Republic*, August 13, 2014. www.newrepublic.com.

Schoenberg, Arnold. *Style and Idea: Selected Writings of Arnold Schoenberg*. Edited by Leonard Stein. Translated by Leo Black. Berkeley: University of California Press, 1975.

Scholz, Trebor. *Uberworked and Underpaid*. Malden, MA: Polity Press, 2017.

———. "The Politics of the Sharing Economy." *Public Seminar*, June 30, 2014. www.publicseminar.org.

———. "Crowdmilking." *Collectivate*, March 9, 2014. www.collectivate.net.

———. "Introduction: Why Does Digital Labor Matter Now?" In *Digital Labor: The Internet as Playground and Factory*, edited by Trebor Scholz, 1–9. New York: Routledge, 2013.

Scott, Darieck. *Extravagant Abjection: Blackness, Power, and Sexuality in the African American Literary Imagination*. New York: New York University Press, 2010.

Scott, Mark. "Study Finds Some Uber and Lyft Drivers Racially Discriminate." *New York Times*, October 31, 2016. www.nytimes.com.

Scott-Heron, Gil. *Small Talk at 125th and Lenox*. New York: Flying Dutchman Records, 1970.

Sedgwick, Eve Kosofsky, and Adam Frank. *Touching Feeling: Affect, Pedagogy, Performativity*. Durham, NC: Duke University Press, 2003.

Shakespeare, William. *The Tempest*. Mineola, NY: Dover, 1999.

Shirky, Clay. *Here Comes Everybody: The Power of Organizing without Organizations*. New York: Penguin, 2008.

Showalter, Elaine. "Our Age of Anxiety." *Chronicle of Higher Education*, April 8, 2013. www.chronicle.com.

Shulevitz, Judith. "It's Payback Time for Women." *New York Times*, January 8, 2016. www.nytimes.com.

Simmel, Georg. *The Philosophy of Money*. Edited by David Frisby. Translated by Tom Bottomore and David Frisby. New York: Routledge, 2004.

Singer, Natasha. "In the Sharing Economy, Workers Find Both Freedom and Uncertainty." *New York Times*, August 16, 2014. www.nytimes.com.

Smith, Aaron, and Janna Anderson. "AI, Robotics, and the Future of Jobs." Pew Research Center's Internet & American Life Project, August 6, 2014. www.pewinternet.org.

Smith, Daniel. "It's Still the 'Age of Anxiety.' Or Is It?" *New York Times*, January 14, 2012. www.opinionator.blogs.nytimes.com.

Smythe, Dallas W. "Communications: Blindspot of Western Marxism." *Canadian Journal of Political and Social Theory* 1, no. 3 (1977): 1–27.

Solman, Paul. "Man vs. Machine: Will Human Workers Become Obsolete?" *PBS NewsHour*, May 24, 2012. www.video.pbs.org.

Srnicek, Nick, and Alex Williams. *Inventing the Future: Postcapitalism and a World without Work*. New York: Verso, 2015.

Stallybrass, Peter, and Allon White. *The Poetics and Politics of Transgression*. Ithaca, NY: Cornell University Press, 1986.

Sterne, Jonathan. "The Historiography of Cyberculture." In *Critical Cyberculture Studies*, edited by David Silver and Adrienne Massanari, 17–28. New York: New York University Press, 2006.

Stewart, James B. "At Google, a Place to Work and Play." *New York Times*, March 16, 2003. www.nytimes.com.

Strauss, Ilana E. "Would a Work-Free World Be So Bad?" *Atlantic*, June 28, 2016. www.theatlantic.com.

Taylor, Mark C. *Confidence Games: Money and Markets in a World without Redemption*. Chicago: University of Chicago Press, 2008.

Terada, Rei. *Feeling in Theory*. Cambridge, MA: Harvard University Press, 2001.

Terranova, Tiziana. "Free Labor: Producing Culture for the Digital Economy." *Social Text* 18, no. 2 (2000): 33–58.

Thompson, Derek. "A World without Work." *Atlantic*, August 2015. www.theatlantic.com.

———. "What Jobs Will the Robots Take?" *Atlantic*, January 23, 2014. www.theatlantic.com.

Timberg, Scott. "Jaron Lanier: The Internet Destroyed the Middle Class." *Salon*, May 12, 2013. www.salon.com.

Toffler, Alvin. *The Third Wave*. New York: Bantam, 1980.

Tompkins, Kyla Wazana. "Intersections of Race, Gender, and Sexuality: Queer of Color Critique." In *The Cambridge Companion to American Gay and Lesbian Literature*, edited by Scott Herring, 173–89. New York: Cambridge University Press, 2015.

"Transcript: Obama's Speech to Congress on Jobs." *New York Times*, September 8, 2011. www.nytimes.com.

Tronti, Mario. "Workers and Capital." *Telos* no. 14 (1972): 25–62.

Tuhkanen, Mikko, ed. *Leo Bersani: Queer Theory and Beyond*. Albany: State University of New York Press, 2014.

Vahamaki, Jussi, and Akseli Virtanen. "Deleuze, Change, History." In *Deleuze and the Social*, edited by Martin Fuglsang and Bent Meier Sørensen, 207–28. Edinburgh: Edinburgh University Press, 2006.

Veblen, Thorstein. *The Theory of the Leisure Class: An Economic Study of Institutions*. New York: Macmillan, 1912.

Wark, McKenzie. "Considerations on a Hacker Manifesto." In *Digital Labor: The Internet as Playground and Factory*, edited by Trebor Scholz, 69–76. New York: Routledge, 2013.

Warner, Michael. *The Trouble with Normal: Sex, Politics, and the Ethics of Queer Life*. Cambridge, MA: Harvard University Press, 2000.

Weber, Max. *The Protestant Ethic and the "Spirit" of Capitalism and Other Writings*. Translated by Peter Baehr and Gordon C. Wells. New York: Penguin, 2002.

Weeks, Kathi. *The Problem with Work: Feminism, Marxism, Antiwork Politics, and Postwork Imaginaries*. Durham, NC: Duke University Press, 2011.

Weiner, Joshua J., and Damon Young. "Queer Bonds." *GLQ: A Journal of Lesbian and Gay Studies* 17, nos. 2–3 (2011): 223–41.

Wellman, Barry. "The Three Ages of Internet Studies: Ten, Five, and Zero Years Ago." *New Media & Society* 6, no. 1 (2004): 123–29.

Wiegman, Robyn, and Elizabeth A. Wilson. "Introduction: Antinormativity's Queer Conventions." *differences* 26, no. 1 (2015): 1–25.

Williams, Raymond. *Politics and Letters: Interviews with New Left Review*. New York: New Left, 1979.

Wiseman, Paul, and Bernard Condon. "Will Smart Machines Create a World without Work?" Associated Press, January 25, 2013. www.bigstory.ap.org.

Wolf, Martin. "Enslave the Robots and Free the Poor." *Financial Times*, February 11, 2014. www.ft.com.

Zelizer, Viviana A. *The Social Meaning of Money*. New York: Basic, 1995.

Žižek, Slavoj. *The Sublime Object of Ideology*. New York: Verso, 1989.

# INDEX

Adorno, Theodor, 46, 75–76

Ahmed, Sara, 2, 6, 168n9; on anxiety, 22–23; on emotion source, 20–22; on normativity, 25–26; on responsibiliza-tion, 37; on willfulness, 108–10

AIDS epidemic, 30

Airbnb, 11, 128

Andreessen, Marc, 93

*Angst* (Kahn), 19

antinormative critique, 27; ethics in, 29

antisocial thesis, 168n9, 169n17; capitalism in, 139–40; critiques of, 37; intersec-tionality with, 4–5; leisure with, 47; Muñoz on, 34–35; privilege in, 4; in queer theory, 2–3

anxiety: ages of, 17–25; Ahmed on, 22–23; from automation, 7, 90, 98–99; boundaries in, 22; conceptualization of, 19–20; as concern, 12–13; from digital technology, 6; fear and, 22; gender in, 23; as health epidemic, 18–19; from leisure-at-work, 81; Ngai on, 22–24; with normative, 6; as political strategy, 2; as popular term, 18; from projection, 22–23; romanticization of, 23–24; of social bond loss, 9; from technological unemployment, 85–86

AP. *See* Associated Press

appropriation, 55

Arvidsson, Adam, 50

Asimov, Isaac, 98

assimilation, 17

Associated Press (AP), 98

*Atlantic*, 100

*Autoaffection* (Clough), 164

automation: abundance with, 83–84; anxi-ety from, 7, 90, 98–99; biology in, 104; capitalist threat of, 105; of cognitive la-bor, 102; concern about, 1–2; economic inequality from, 92–93; headlines on, 83–84; job sectors at risk for, 100; labor distinctions with, 104–5; Lanier on, 94–96; in nature, 96; precarity from, 151–52; of professional white-collar labor, 100–101; state role in, 89–94; work safe from, 91. *See also* robots

*Baffler*, 154

Barad, Karen, 21

barebacking, 29

Barlow, David H., 23

Battelle, John, 39

Baym, Nancy, 58

Beller, Jonathan, 48

belonging, 134

Bersani, Leo, 3, 6; on homosexuality, 32–33; on narcissism, 35; on socialization, 30–31

Best, Stephen, 15

Bezos, Jeff, 127

biology: in automation, 104; in labor, 103–4

Black Lives Matter movement, 4

Bloch, Ernst, 22

blue-collar labor, 111–12

Bogost, Ian, 53

Botsman, Rachel, 122

boundaries: in anxiety, 22; defining of, 24; to emotions, 21

## ABOUT THE AUTHOR

Greg Goldberg is Assistant Professor in the Department of Sociology at Wesleyan University, and Affiliated Fellow at Yale University's Information Society Project. His work has appeared in *New Media & Society*, *Social Media & Society*, *WSQ*, *and ephemera*, and in the edited collections *The Affective Turn* and *Rethinking the Innovation Economy*.

Printed in the United States
By Bookmasters